Cities and Econ

Cities and Economies explores the complex and subtle connections between cities and economies. The rise of the merchant city, the development of the industrial city and the creation of the service-dominated urban economy are explored, along with economic globalization and its effects on cities in both developed and developing economies. This book provides a thorough examination of the role of the city in shaping economic processes and explains the different effects that economies have on cities. It provides an invaluable and unrivaled guide to the relationship between urban structure and economic processes as they compare and contrast across the world.

The authors examine the complex relationships between the city and the economy in historical and global contexts, as well as evaluating the role of world cities, the economic impacts of mega cities, and the role of the state in shaping urban economic policies. They focus on the ways in which cities have led, and at the same time adapted to, economic shifts. Large cities are viewed as the centers of regional and national economies, while a small number are defined by their centrality in the global economy.

The book examines:

- Key ideas and concepts on the economic aspects of urban change.
- The changing nature of urban economies and their relationships with changes at the national and global levels.
- Current economic issues and policies of large cities around the world.
- The links between globalization and economic changes in cities and the growing competitions between them.

Cities and Economies uses case studies, photographs, and maps of the US, Western Europe and Asia. Written in a clear and accessible style, the book answers some fundamental questions about the economic role of cities. *Cities and Economies* is an essential text for students of Geography, Economics, Sociology, Urban Studies, and Urban Planning.

Yeong-Hyun Kim is Associate Professor in Geography at Ohio University, USA. Her teaching and research specialty includes globalization, urban political economy, development, and Asia. She is currently working on migrant labor in Seoul, South Korea.

John Rennie Short is Professor of Geography and Public Policy at the University of Maryland, Baltimore County, USA. He has published 30 books and numerous articles and is recognized as an international authority on the study of cities.

Routledge critical introductions to urbanism and the city

Edited by Malcolm Miles, University of Plymouth, UK
and John Rennie Short, University of Maryland, USA

International Advisory Board:

Franco Bianchini	Jane Rendell
Kim Dovey	Saskia Sassen
Stephen Graham	David Sibley
Tim Hall	Erik Swyngedouw
Phil Hubbard	Elizabeth Wilson
Peter Marcuse	

The series is designed to allow undergraduate readers to make sense of, and find a critical way into, urbanism. It will:

- cover a broad range of themes
- introduce key ideas and sources
- allow the author to articulate her/his own position
- introduce complex arguments clearly and accessibly
- bridge disciplines, and theory and practice
- be affordable and well designed

The series covers social, political, economic, cultural and spatial concerns. It will appeal to students in architecture, cultural studies, geography, popular culture, sociology, urban studies, urban planning. It will be trans-disciplinary. Firmly situated in the present, it also introduces material from the cities of modernity and post-modernity.

Published:
Cities and Consumption – Mark Jayne
Cities and Cultures – Michael Miles
Cities and Nature – Lisa Benton-Short and John Rennie Short
Cities and Economies – Yeong-Hyun Kim and John Rennie Short

Forthcoming:
Cities and Cinema – Barbara Mennel
Cities, Politics and Power – Simon Parker
Urban Erotics – David Bell and John Binnie
Children, Youth and the City – Kathrin Hörshelmann and Lorraine van Blerk
Cities and Gender – Helen Jarvis, Jonathan Cloke and Paula Kantor

Cities and Economies

Yeong-Hyun Kim and
John Rennie Short

 Routledge
Taylor & Francis Group

LONDON AND NEW YORK

First published 2008
by Routledge
2 Park Square, Milton Park, Abingdon, Oxon, OX14 4RN

Simultaneously published in the USA and Canada
by Routledge
270 Madison Avenue, New York, NY 10016

Routledge is an imprint of the Taylor & Francis Group, an informa business

© 2008 Yeong-Hyun Kim and John Rennie Short

Typeset in Times New Roman and Futura by
Keystroke, 28 High Street, Tettenhall, Wolverhampton
Printed and bound in Great Britain by
Antony Rowe Ltd, Chippenham, Wiltshire

British Library Cataloguing in Publication Data
A catalogue record for this book is available from the British Library

Library of Congress Cataloging in Publication Data
Kim, Yeong-Hyun.
Cities and economies / by Yeong-Hyun Kim and John Rennie Short – 1st ed.
p. cm.
Includes bibliographical references and index
1. Urban economics. 2. Sociology, urban. 3. Cities and towns–History.
4. Globalization. I. Short, John. II. Title.
H321.K554 2008
330.9173'2–dc22   2007025730

ISBN10: 0–415–36573–2 (hbk)
ISBN10: 0–415–36573–0 (pbk)
ISBN10: 0–203–01827–3 (ebk)

ISBN13: 978–0–415–36573–4 (hbk)
ISBN13: 978–0–415–36574–1 (pbk)
ISBN13: 978–0–203–01827–9 (ebk)

Contents

List of illustrations   vii

List of case studies   xi

Acknowledgements   xiii

1 Cities and economies   1
   Urban growth and decline   2
   Large cities in the national and global economies   4
   Urban economic changes in the era of globalization   7

Part One: Urban growth and decline in economic shifts   **11**

2 Mercantile cities and European colonialism   13
   Merchant capitalism and the growth of towns   14
   Merchant capitalism, colonialism and global urban networks   20
   Cartographic representations of the merchant city   25

3 The rise and fall of industrial cities   28
   The industrial city and social conflict   32
   The planned city   34
   The Fordist city   37

4 Service industries and metropolitan economies   48
   The growth of the service sector   50
   The role of the public sector   56
   Global shift of services   57

**Part Two: The global economy and world cities in
developed countries**   **63**

5 Globalization and world cities   65
World cities in urban studies   66
Command and control centers of the global economy   72
The global urban hierarchy   74

6 Globalization and globalizing cities   80
Urban competitiveness   82
Politics of world-city status   89
Globalizing cities and globalized urbanism   93

7 New solutions for old economies   97
The cultural economy   98
The creative economy   104

**Part Three: The national economy and capital cities
in developing countries**   **115**

8 Third world cities   117
Third world cities in urban studies   118
Mega cities of the developing world   124
Primate cities   134

9 World city projects for national capitals   140
National economic development and capital cities   141
Graduating from third world city status   146

10 Globalizing islands in developing countries   156
Inequality in globalizing third world cities   157
High-tech enclaves in Bangalore   162
Ethnic enclaves in Seoul   165

References   173

Index   191

Illustrations

Tables

3.1	The four Kondratieff cycles	29
4.1	The 20 largest cities in the US, 1850–2000	52
5.1	Friedmann's world city hierarchy	67
5.2	American cities with highest percentages of foreign-born residents	70
5.3	Changes in the percentage of foreign-born residents in selected US cities	71
5.4	Top ten cities in Global 500	73
5.5	International passenger traffic, April 2005–March 2006	75
5.6	The world's ten most "powerful and prestigious" cities	76
5.7	World cities in a culturally globalizing world	77
6.1	Population decline in large industrial cities in Ohio, 1900–2000	85
6.2	Ranks of industrial cities in Ohio by population of the 100 largest US urban places, 1810–2000	86
6.3	Atlanta's New Century Economic Development Plan	93
7.1	America's most creative cities	107
7.2	Toronto's recommended strategies for a creative city	112
8.1	The world's mega cities with 10 million people or more	125
8.2	A typology of economic and population changes in cities	128
8.3	Urban and slum population in the world, 2001	129
8.4	Types of economic activity	130
8.5	Examples of primate cities	135
8.6	Most and least desirable cities for expatriates, 2006	138
9.1	New capital projects in the developing world since 1950	143
9.2	The world's tallest buildings, 2006	147
9.3	Population changes in Seoul, 1949–2005	150
9.4	Host and bid cities of the Summer Olympics, 1980–2012	154

Figures

2.1	Periodic market systems	15
2.2	Seville in *Civitates Orbis Terrarum*	26
4.1	Geographical shift of the 20 largest cities in the US, 1900 and 2000	53
5.1	Regional world cities and their spheres of influence	69
5.2	GaWC's global network connectivity	78
6.1	Atlanta toward a world-class city	92
7.1	Creative City, Yokohama	110
8.1	Satellite–metropole relations in the world economy	123
8.2	Location of mega cities in 2005	127
9.1	Jo'burg: an African world city and a World Cup city	149
9.2	The Seoul Metropolitan Area	151
10.1	Major sources of foreign-born IT workforce in Silicon Valley, 2000 and 2005	167
10.2	Migrant workforce in South Korea	170

Plates

1.1	Changed economic fortune: Main Street, Over-the-Rhine, Cincinnati, Ohio	4
2.1	Florence on the banks of the Arno River	17
2.2	Portuguese colonial legacy in Chennai, India	23
3.1	Haussmann's Paris	36
3.2	Industrial park in Ho Chi Minh City, Vietnam	47
5.1	Hong Kong's claim: Asia's world city	68
6.1	Glasgow: Scotland with Style campaign	91
7.1	City for the Arts, Arts for the City, San Francisco, California	101
7.2	Contemporary Arts Center, Cincinnati, Ohio	103
7.3	Not so aesthetic for creative people: Riverside Plaza in Minneapolis, Minnesota	108
8.1	Deteriorating housing conditions in Kolkata (Calcutta), India	128
8.2	Water pollution in Delhi, India	131
9.1	The Petronas Towers in Kuala Lumpur	147
10.1	Learn to develop new software in Chennai, India	163
10.2	Ethnic business center in Outer Seoul, South Korea	171

Boxes

6.1 American city mayors' commitment to economic development 83
7.1 Cultural development for traditional industrial cities 102
7.2 History of CreativeTampaBay, Florida 106
7.3 Osaka's creative city initiative 111

Case studies

2.1 Madras/Chennai 22
3.1 Locational theories of industrial cities 31
3.2 Contrasting urban fortunes 41
4.1 The twenty largest cities in the US, 1850–2000 51
5.1 American cities, cosmopolitan cities 70
6.1 Urban decline in Ohio 84
7.1 Creative cities in Japan 109
8.1 Mega cities: decoding the chaos 132
8.2 Third world cities in the league table of quality of life 137
9.1 Growth pole theory 145
9.2 Johannesburg: building an African world-class city 148
10.1 Gated communities in Trinidad 161
10.2 Bangalore: India's Silicon Valley or Silicon Valley's India? 166

Acknowledgements

We would like to thank the series co-editor Malcolm Miles as well as Andrew Mould and Jennifer Page at Routledge. We would also like to thank Kenny Ling for his help with graphics on Figures 4.1 and 8.2.

We are grateful to the following for granting permission to reproduce copyright material: City of Atlanta, the US for Figure 6.1 Atlanta toward a world-class city; City of Yokohama, Japan for Figure 7.1 Creative City, Yokohama; and City of Johannesburg, South Africa for Figure 9.1 Jo'burg.

While the authors and publisher have made every effort to contact copyright holders of material used in this book, in a few cases we were unable to do so. All photographs are by Yeong-Hyun Kim.

1 Cities and economies

Learning objectives

- To look at urban growth/decline in the historical and global contexts
- To think about the causes and outcomes of urban growth/decline
- To understand the significance of large urban agglomerations in the national and global economies
- To describe the difference between world cities and third world cities

The relationship between cities and economies has been examined by many disciplines at various sites using different research methods. Most of the studies revolve around these fundamental questions:

- Why do cities exist?
- How did early cities emerge?
- Why do some cities grow while others stagnate or decay?
- How can government policies help urban economies develop, or prevent their decline?
- Who benefits the most, and the least, from urban economic growth?
- What roles do cities, particularly large ones, play in regional and national economic development?
- What impact does economic globalization have on cities?

Different disciplines traditionally ask, and answer, the questions differently. Urban economists tend to look for the underlying causes of the spatial concentration of economic activities. In contrast, urban geographers tend to focus more on the outcomes of such concentration that include population growth, residential segregation and suburbanization. Finally, urban policy-makers concentrate more on problems caused by agglomeration, such as traffic congestion and crime.

In this book we focus on three interrelated topics, which will echo throughout the rest of the text: urban growth and decline; large cities in the national and global economies; and urban economic changes in the era of globalization.

Urban growth and decline

Cities grow or decline for different reasons at different times. What causes certain cities to grow at particular times, or in general, remains one of the most frequently asked questions in urban studies. Changes in urban fortune are easy to explain for some cities, say, if their economies have been established around one or two very prominent industries, such as mining. The same is true for cities that owe much of their economic growth to geographical location, transportation routes, abundant natural resources, political functions or tourist sites. Yet, sometimes, urban economy appears arbitrary, as some cities flourish, and other cities in similar situations suffer economic decline.

How do we explain different development paths that different cities have taken? Is geographical location still key to a city's growth or decline? Or its government's economic policies? How do local traditions – such as the relative value placed on the entrepreneurial spirit – affect it? And what about the human and cultural capital that cities have built? Finally, do pure luck and happenstance play a role in urban growth and decline? Urban scholars have attempted to identify major sources of urban economic change that could apply to a wide range of cities across times, cultures and the world. They start by asking why cities exist.

The concept of "increasing returns" to the scale is commonly used to explain the geographical agglomeration of economic activities in cities (Fujita *et al.*, 1999; Henderson, 1988). Simply put, cities emerge and grow when increasing returns exceed transportation costs. However, modeling the increasing returns to spatial concentration has proved problematic for "perfect competition." Most urban geographers would agree on the notion that spatial concentration itself creates a favorable economic environment by providing the technological spillovers and socio-cultural networks that support continued concentration. Meanwhile, many urban economists like to dissect the initial conditions that break the spatial equilibrium in the first place and lead to agglomeration. In other words, some examine the effects of "linkages," "externalities," "multiplier effects" and "circular cumulative causations" in urban economy and space, while others focus on the initial causes of these economic processes.

Those who consider increasing returns to be the most critical factor in the emergence of cities may find little to agree on within traditional theories of urban origin. Such theories point towards rather non-economic factors, such as religious causes or defensive needs (Childe, 1950; Mumford, 1961). In his seminal book *The City in History*, Mumford (1961: 10) links the first germ of the city to "the ceremonial meeting place that serves as the goal for pilgrimage: a site to which family or clan groups are drawn back, at seasonable intervals, because it concentrates, in addition to any natural advantages it may have, certain 'spiritual' or supernatural powers, powers of higher potency and greater duration, of wider cosmic significance, than the ordinary processes of life."

Not all those who see economic factors as critical to the emergence of early cities use highly sophisticated equations and complex graphs to prove their point. Jane Jacobs' works utilize a descriptive method to explain economic changes in cities. In *The Economy of Cities* (1969), she uses an imaginary city, named New Obsidian and located on the Anatolian plateau of Turkey, to develop a theory in which early cities support their surrounding rural areas. This disputes the commonly accepted notion that cities build upon a rural economic base. Urban economic growth, notes Jacobs (1969: 49), takes place by "adding new kinds of work to the existing." In the case of New Obsidian, animal domestication was added to the obsidian trading, while many modern cities, such as Los Angeles and Tokyo, have been successful in adding new export works to their local economies.

This book does not attempt to resolve how the returns to spatial concentration should be modeled or under what conditions small differences among locations snowball into larger differences over time. We gladly leave that task to urban economists, particularly those in agglomeration economics or geographical economics (Brakman *et al.*, 2001; Fujita and Thisse, 2002). Nor do we seek to build a new theory of urban origin. Outstanding scholarship has already been established on the dawn of cities. Instead, we aim to examine the economic factors that have caused, and been caused by, the growth and decline of large cities over time.

During the mercantile era, urban economic growth was closely linked to international trade, as port cities at junctures of profitable trade routes grew in size and prominence. After the Industrial Revolution, manufacturing centers gained populations rapidly and subsequently positioned themselves on the leading edge of social and economic change. In the past few decades, however, many such great old cities have experienced significant decline. Some industrial cities have successfully diversified their economic base by promoting the growth of high-value added service sectors, yet some have not. Plate 1.1 presents a good example of those unfortunate ones. The Over-the-Rhine area in Cincinnati, Ohio, was a densely populated, economically vibrant immigrant neighborhood in the mid-nineteenth century, when Cincinnati was a leading manufacturing city and trading center in the US. With the decline of canal transportation and manufacturing, this once vibrant community has suffered all the ill effects of urban decline. The city government's Over-the-Rhine Comprehensive Plan might help turn around the economic fortune of this area in the future, but it continues to lose population. This book seeks to look at both causes and consequences of changed urban fortunes and government policies to spark and sustain economic growth.

Large cities in the national and global economies

Large cities are viewed as a mixed blessing to human society. Given the extent of urban problems in the developed and developing worlds alike, we routinely associate large cities with some distressing and even horrifying images. Literary portrayals of the urban experience by writers such as Charles Dickens and Franz Kafka have bolstered the problematic image of modern cities, particularly large

Plate 1.1   Changed economic fortune: Main Street, Over-the-Rhine, Cincinnati, Ohio

industrial ones (Alter, 2005; Williams, 1973). A reader of their novels might conclude that modern cities overflow with poverty, crime and traffic congestion, while lacking in meaningful interaction, a sense of community and sanity. This tilt toward troubled city depictions, rather than vibrant and cheerful ones, is even more pronounced in scholarly works. The urban poor receive far more academic attention than the urban rich. While there is nothing inherently "wrong" with social scientists investigating socio-economic problems in poor neighborhoods, the fact remains that far less academic research has been conducted on urban wealth than on urban poverty.

As such, cities with a host of social and economic problems, as most large ones have, are often seen not as generators of, but as impediments to, the economic growth of their nations (Kasarda and Parnell, 1993). Indeed, many large cities in low-income countries unmistakably demonstrate serious diseconomies of large-scale urbanization. Such countries may discuss limiting urban growth to promote national economic development, although the practicality of such a policy remains debatable (Laquian, 2005).

We argue that in large cities, the good outweighs the bad – even in developing countries. According to Scott (2006: 99), the long-term benefits to urban growth in less developed countries appear to outweigh the costs in almost every case. Polarization reversal policies, once vigorously advocated as a measure for national economic development, are no longer recommended by either international development agencies or development scholars. In *City Economics*, O'Flaherty (2005: 2) notes that urban difficulties "arise out of the same geographical propinquity that makes cities work. You can't have one without at least encountering the other – just as you can't water ski without getting wet." Despite the diseconomies and other problems of urban growth, large cities are growth machines for the national economy and, in recent decades, the global economy.

Technological advancement in telecommunications is another factor that has threatened the very existence of large cities, particularly in highly developed countries. Futurists like Nicholas Negroponte (1995) and John Naisbitt (1994) projected that with the coming of the digital era, the need to live and work in large cities will disappear, and, with it, urban overcrowding. Information and telecommunication technologies may have affected economic change in some cities, yet the evident attraction of large cities and their business districts remains. Every weekday, for example, more than 1.3 million people commute to work in Manhattan, nearly doubling its resident population. A *New York Times* article (McGeehan, 2006) claims that the recent resurgence of Manhattan as a site for corporate headquarters, despite its skyrocketing rent and horrific traffic jams, can be attributed in part to technological innovations that allow corporate executives to stay in touch at a distance with their staff.

Large cities are sites of high efficiency and productivity in a capitalist economy. They have achieved spectacular growth through trade, industrialization, modernization, nation building, (im)migration and now globalization. The significance of large cities in national development and, in general, world development has been noted in a variety of academic writings. O'Flaherty (2005: 1) states in plain terms that "Life without cities would be poorer than it is now – not just for city residents but also for anyone who consumes the products and services that are developed and produced in cities, and anyone whose wages are higher because he or she could migrate to a city. These categories include just about everyone in the world today." Savitch (1996: 46) claims that "urban centers have always led in the creation of national wealth, and many now occupy a special place in the global era." His appreciation for large cities continues in a co-authored book (Savitch and Kantor, 2002: 3): "Cities are the crucibles through which radical experiments become convention. They are concentrated environments in which people adapt and their resilience is tested. They are the world's incubators of innovation – made possible by critical mass, diversity, and rich interaction. And cities have steadily grown over the centuries to fulfill that role."

With the growing popularity of the global–local framework in social science researches, urban changes in large cities are now routinely examined in a global context. The impact of globalization on cities appears in areas of economic restructuring, downtown redevelopment, neo-liberal urban politics, social polarization and multicultural urban lifestyles, among others. John Friedmann's (1986) world city hypothesis suggests that urban change be explained with reference to a worldwide economic process, inspiring many urban scholars to assume more global perspectives. Prior to Friedmann's work, the global–local nexus rarely came up while investigating urban issues in the developed world, particularly North America, where urban economic dynamics were seldom linked to forces beyond their national borders, such as international networks of goods, capital or migrants (Barlow and Slack, 1985; Davis, 2005).

In contrast, following Andre Gunder Frank's metropole–satellite model (1966), third world urbanization has always been contextualized in international political and economic processes (Smith, 1996; Timberlake, 1985). External forces, mainly relating to colonialism and post-colonialism, have long appeared in the studies of large cities in the developing world.

However, the recent explosion of writings on world cities does not include many case studies of third world cities, as few of them seem to play a "command-and-control" role in economic globalization. Ironically, this recent wave of world-cities research has not built upon the existing studies of third world urbanization; indeed, it largely ignores the global aspect of third world cities. In this book, we intend to examine how large cities, both world cities in the developed world and third world cities in the developing world, have transformed and how their relationships with national and global economies have restructured over time.

Urban economic changes in the era of globalization

In *The Restless Urban Landscape*, Paul Knox (1993) and other contributors illustrate the constant formation and reformation of urban landscapes in response to the imperatives and contractions inherent to the dynamics of a capitalist economy and society. The dynamics of the capitalist system might be the ultimate answer to the question, "What causes restless change in the urban economy?" The terms like "capitalist city" (Smith and Feagin, 1987), "world city network" (Taylor, 2004) and "global (or transnational) urbanism" (Smith, 2001) suggest that comparable urban economic structures, urban landscapes and urban lifestyles exist throughout the world, and that capitalism explains the similarities in global urban changes. However, some identifiable forces independent of capitalism also affect urban economic changes. To consider these forces, we must step outside the familiar territory of Anglo-American cities.

Given the vast global diversity of politics, geography, culture and history, identifying universal causes of urban economic change can present a challenge. Certainly technological advancement has transformed every urban economy in the contemporary world (Castells, 2000). But beyond this universal cause, few forces have triggered parallel urban changes in both the developing and the developed world.

Urban scholars generally agree that the combined forces of deindustrialization and suburbanization, along with globalization, have played pivotal roles in recent economic changes in American cities. These processes also help to explain recent changes in the economies of many European cities. However, the intergovernmental supports and local political cultures in Europe differ greatly from those in American cities. In addition, waves of immigration have altered, and continue to alter, the economic structures of many Western cities. Additional forces could be added to the list of deindustrialization, suburbanization and globalization, but the combined role of these three factors remains key to urban economic change, both in North America and in Western Europe.

It is not clear, however, how current urban experiences in Africa, Asia and Latin America can be generalized into a few processes. Some debate the relevance of even attempting to generalize their vast experiences (Potter and Lloyd-Evans, 1998). Pointing to poverty as a pronounced economic process in third world cities may prove unproductive, since the concept of wealth is not used (at least, not in urban studies) as such to analyze recent change in urban American or European economies. Along with poverty, industrialization and modernization still dominate research frameworks for urban changes in third world cities. These concepts have been linked to processes of rural–urban migration and over-urbanization, reinforcing the nightmarish image of third world cities with regards to urban governance and planning. The sobering reality is that, while there have been plenty of reports, including the United Nations Human Settlements Programme publications (UN-HABITAT reports), of economic hardships and slums in third world cities, few efforts have been made to theorize their urban economic conditions (Short, 2006b).

The concept of post-colonial urbanism offers a rare attempt to link theoretical perspectives to historical and empirical accounts (Bishop *et al.*, 2003). In her book *Ordinary Cities*, Jennifer Robinson (2006) calls for a post-colonial urban studies in which third world cities are not judged by what they lack or what their governments fail to do. Instead, they should be viewed as a cosmopolitan source of urban theory. While Western cities have been the almost exclusive sites for the production of key concepts and theories in urban studies, a post-colonial understanding of cities will draw on more diversified experiences and inspirations.

In this book, we examine urban economic changes in the developing world as well as the developed world, although the relative lack of research on third world cities limits the scope and depth of our work. We also examine the relationship between cities and economies in the historical and global contexts. The text will focus, in particular, on the ways in which large cities have led, and at the same time adapted to, economic shifts at the national and international levels. Three topics – urban growth and decline; large cities in national and global economies; and the impact of globalization on cities – interweave throughout the book.

Part One, "Urban growth and decline in economic shifts," examines different economic activities that have spurred urban growth and decline in history, including international trade for mercantile cities, manufacturing for industrial cities and service industries for large metropolises in the contemporary world.

Part Two, "The global economy and world cities in developed countries," focuses on how large cities in developed countries have led, and adapted to, a globalizing world economy. Some of these cities have been identified as world cities, and many others strive for world city status. The three chapters in this section examine the roles that world cities play in a global economy, how globalization escalates competition between cities, policies used by city governments to improve their cities' competitiveness in international market places, and suggested solutions for declining urban economies.

We devote Part Three, "The national economy and capital cities in developing countries," to large cities in developing countries, with particular focus on the links between capital cities and their national economic development. It is not always the case that large cities in the developed world are global, while those in the developing world remain national. However, this book focuses on the globally linked aspects of the former in Part Two, and on the nationally centered aspects of the latter in Part Three. The three chapters of Part Three examine how third world cities and their economic problems have been researched in the social sciences, how national governments of developing countries have restructured their capital cities through world city projects, and how globalization has effected inequality in large, globalizing cities of the developing world.

Further reading

Jacobs, Jane, 1969, *The Economy of Cities*, New York: Random House. The late Urban Studies legend Jane Jacobs writes about how certain cities continue to grow by adding new work to the old. It is a must-read book for those who are interested in urban growth/decline.

O'Flaherty, Brendan, 2005, *City Economics*, Cambridge: Harvard University Press. This introductory book discusses a wide range of urban economic issues and public policies, including housing, mass transit and crime, in an accessible and engaging style.

Savitch, H.V. and Paul Kantor, 2002, *Cities in the International Marketplace: The Political Economy of Urban Development in North America and Western Europe*, Princeton: Princeton University Press. This book examines comparatively development policies at ten cities in North America and Western Europe. The authors argue that urban economic development policies are formulated at the juncture of local politics and the international market place.

Short, John Rennie and Yeong-Hyun Kim, 1999, *Globalization and the City*, New York: Longman. The authors look at the impact of economic, cultural and political globalization on world cities.

Part One

Urban growth and decline in economic shifts

In Part One we take a broad historical sweep to consider the long cycles of urban economic growth and decline. We review different economic activities that have spurred urban growth/decline in history, including international trade for mercantile cities, manufacturing for industrial cities and service industries for large metropolitan regions in the contemporary world.

Chapter 2 looks at the rise of merchant cities in the context of the development of long-distance trade. Networks of trading cities at both the regional and global scale are identified. The rise and fall of merchant cities is linked to widening circuits of commodity exchange and the development of European colonialism. Merchant cities formed one of the earliest global urban networks.

Chapter 3 looks at the rise of the industrial city. Embodying the connection between urbanization and industrialization, the industrial cities in the developed world transformed the landscape, reshaped social relations and introduced a new type of city and a new form of urban economic growth. The factory, sharp class relations and the rise of an organized working class were all incubated in the industrial city. We begin our discussion by considering Manchester in northern England and go on to look at the rise and fall of the industrial city in the developed world. The changing role of manufacturing is assessed as we examine the deindustrialization in the West and the growth of industrial districts in selected parts of the developing world.

In Chapter 4 we look at the shift from manufacturing to service employment and especially the reasons behind the agglomeration of advanced producer services in a narrow range of global and globalizing cities. Selected cities are the control and command centers of a global economy, the urban sites of economic globalization. While most developed economies have witnessed the rise of mercantile cities, industrial cities and service-sector-led metropolises in historical sequence, many developing economies have taken different paths in their urban development histories. We examine the links between cities and economies in developing countries as well as developed countries.

Part One provides the broad historical perspective that informs the more detailed discussion in Parts Two and Three.

2 Mercantile cities and European colonialism

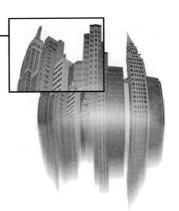

Learning objectives

- **To explore the development of merchant cities**
- **To understand central place theory**
- **To understand the links between colonialism, merchant capitalism and city growth**

When Ferdinand Magellan sailed into the island of the Philippine archipelago in March 1521 he was surprised to find silk, porcelain and other artifacts of Chinese material culture. The European sailors were presented with evidence that other trading empires had visited these islands before them. Magellan was stumbling across the vestigial remains of the once extensive Chinese trading empire. It had begun with the ascension of the new emperor Zhu Di in 1402. The new emperor had a vision of vast global commercial empire. He named one of his eunuchs, Cheng Ho, as admiral. Cheng Ho marshaled the immense power and vast resources of the Chinese empire to assemble a fleet of 1,500 ships and thousands of sailors along the banks of the Yangtze at Nanking. It was the largest single fleet in human history until the British Royal Navy of the nineteenth century. On the first voyage in 1405 the Treasure Fleet, as it was known, sailed to India. Later voyages sailed as far as Africa and throughout Southeast Asia and the Pacific. The fleet established trading and diplomatic relations. But Zhu Di died in 1424 and the new ruler, deeply suspicious, cancelled future voyages. However, this new emperor lived for only a few years and his successor, the grandson of Zhu Di, restored the Treasure Fleet. In 1431 the seventh voyage consisted of over 300 ships and 27,000 men, again commanded by Cheng Ho. This would prove to be the peak of Chinese maritime exploration. The Treasure Fleet was soon mothballed, ships were destroyed and China began its long withdrawal from the external world. By 1500 it was illegal to go to sea in large ships. From thereon it was to be the Europeans who would dominate the maritime trading routes. Fraying fragments of silk, half-remembered Chinese phrases and the odd piece of distinctive blue plate were all that were left of China's brief tenure as a maritime power by the time Magellan sailed into the Philippines. The Treasure Fleet reminds us, however, of the contingencies of global dominance. If Cheng Ho had been the first of a long line of Chinese maritime explorers, rather than standing alone, we would have a very different geopolitics. As it was, the Chinese disappeared from ocean-going voyages. By a quirk of history, it was the smaller, more technologically backward European nations that were to dominate the stage.

Merchant capitalism and the growth of towns

The forces that propelled Columbus across the Atlantic and Magellan across the Pacific had been building for centuries. Mercantile capitalism developed out of the economic growth in medieval Europe. As early as the twelfth century identifiable markets with variable prices were in operation. The earliest markets were local, periodic and held in towns and cities. Farmers selling their produce, merchants hawking goods and peddlers selling their crafts would move so markets were distributed in time as well as space. There was an order to the periodicity. Figure 2.1 shows a four-town market arrangement. While the arrangement in Figure 2.1a is designed to minimize distance traveled Figure 2.1b is structured to maximize demand. In the provision of most goods and services travel adds costs; so minimizing travel decreases costs. But for some more expensive goods and services demand is less frequent and so greater consideration is given to maximizing demand; this is done by making sure that markets close together in space are separated in time. As economic development increased the level of demand as well as the amount of goods and services markets in certain cities became fixed in space. Fernand Braudel (1982) provides copious examples of the urban market system in operation in early modern Europe.

Walter Christaller (1966) provided a theory of how such market towns are distributed across space. He assumed a flat, homogeneous plain of even demand, and developed the ideas of the *range* and *threshold* of goods and services. The range of goods and services is the distance people are prepared to travel to purchase the good or service. In the medieval era and early modern period the distance was limited by the cost of transport and the time taken to make journeys. The threshold of a good or service is the minimum population necessary to support a continued supply of the good or service. Lower-order goods and services have a small threshold and a restricted range. Higher-order goods and services

a) minimizing distance

b) maximizing demand

**Figure 2.1
Periodic market
systems**

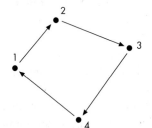

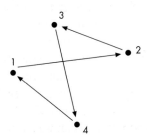

have large thresholds and extensive ranges. Christaller suggested that market towns (he called them central places) would be distributed as a hierarchy in which a large number of small places would provide lower-order goods and service and a small number of larger places would provide higher-order goods and services. Successive levels of the hierarchy would have higher-order goods and services. He identified a number of different hierarchies: a market optimizing hierarchy would minimize the travel of buyers; in the traffic-optimizing hierarchy routes pass through lower- and higher-order places; and in the administration-optimizing hierarchy each lower-order place is completely within the boundary of a higher-order place. This is a robust model of how markets are held together in local and regional space, and provides a useful picture of the early market towns of Europe. Christaller applied his model to southern Germany in the 1930s and found a close fit, which is a testament to the enduring legacy of the medieval market town system. But central place theory breaks down when manufacturing and production replaces the simple buying and selling of goods and services as the primary determinant of town location.

By the early fourteenth century trading connections were lengthening. The demand for higher-order goods, luxury goods like silk, precious metals and spices, was extending tentacles of trade across the globe. By around 1300 we can discern a global urban network of trading flow and commercial connections. Lisa Jardine (1996) shows how the expansion of intellectual and cultural life in the European Renaissance was an expression of an expansion of commerce fueled by conspicuous consumption of luxury goods. She highlights the opulence and materiality of Renaissance cultural expression and argues convincingly that commercial considerations underpinned intellectual advances and cultural movements.

Janet Abu-Lughod (1989) identifies a transcontinental archipelago of cities as early as the fourteenth century that includes the European cities of Bruges, Genoa and Venice, cities in the Arab world including Constantinople, Damascus, Aden and Tabriz, and Peking, Canton and Malacca in the Far East. Not all cities were connected with all the others; rather there was an overlap between regional networks. The early European network centered on Bruges, Genoa and Venice. The cities became important centers where trade was conducted within and between different merchant trading companies. The Saminiati merchant trading company, for example, was established in Leghorn and Florence but eventually had representatives in Cadiz, Lisbon, Paris, Frankfurt, Lille, London, Amsterdam, Hamburg, and Vienna. It used Dutch and English boats and its recorded transactions cover Algiers, Constantinople, Genoa, Marseille and Messina.

The development of world trade and urban development went hand in hand. Successful long-distance trade implied and engendered the creation of a money economy, the emergence of a merchant class and the growth of urban trading centers. Money became a medium of exchange, and although currencies varied widely they were exchangeable, so money slowly displaced barter. A monetarized society meant a quantifiable society. Different goods and services could be more easily traded. The growth of the money economy was part of what Crosby (1997) describes as the "quantification" of Western society between 1250 and 1600 that was part cause and part effect of the success of European commercial imperialism.

Merchants became the driving force of a modernizing society and played a crucial role in the creation of a new, more commercial urban culture. Long-distance trade was risky; so merchants needed to pool their resources to spread the risk. Trading companies and other co-operative business ventures were established. Banks were created to finance the credit lines needed for the long-term investment horizons of far-distance trade. Collective merchant interests crystallized into the commercial propulsions of the city states. In reality it was never quite so simple or so mechanical as indicated here, but over the long term merchant capitalism laid the basis for the modern urban world.

Prior to 1400 there were two main trading networks in Europe. In the north there was the Hanseatic group of almost 80 cities. The name comes from the word *Hanse* (group of merchants). There were the hometowns of the Hanse merchants, Cologne, Dortmund, Lubeck, Danzig and Visby; cities where they had rights to trade, including Bruges, London, Bergen, and Novgorod; as well as cities where they had a presence but no special privileges. The cities were nodes in the trading flows of the primary products of the region, including wood, furs, rye and wheat from the east in exchange for cloth, wine and salt from the west. Lubeck was the linchpin of this east–west trade. The Hanseatic League was a regional merchant trading association that linked cities in a series of trade flows (Dollinger, 1970; Taylor, 2004). In the south trade centered on the Mediterranean. Venice, Florence and Genoa were primary centers whose trading links extended around the margins of the sea as well as further inland (Plate 2.1). From 1350 to the 1700s there was a rise and fall in the fortunes of the dominant cities as Bruges, Venice and then Amsterdam superseded one another in commercial dominance. Let us briefly consider each city.

Bruges initially developed as a trading center for cloth. Its location, with easy sea access, was ideal for the increasing trade with England, the Hanseatic League cities and the trading centers in the Mediterranean. English wool was shipped to Flanders and made into cloth in Bruges and other Flemish towns, such as Ypres and Ghent. Bruges became the trading center where London and Genoese merchants came to buy cloth. By the fourteenth century the city was the largest cloth market in Europe and merchants were also bringing their own goods to trade. It became the mart of Europe, where oranges from Spain, spices from the Orient, furs from Russia and silk and armory from Italy were available alongside the highly prized cloth. The city was essentially run by the merchant class, who created urban public services of law and order, health, excise and tax collection. Revenues from taxes paid for expensive and extensive public works, such as the provision of water and the building of the market and town halls, as well as fortifications and waterways that linked the city to the sea. It was cosmopolitan, with foreign merchants taking up permanent residence in the city. A commercial society was intertwined with religion. Holidays revolved around religious celebrations and wealth enabled the merchants to decorate churches with stained glass and elaborate altarpieces. The city's material culture embodied conspicuous

Plate 2.1 Florence on the banks of the Arno River

consumption in collective enterprises, urban spectaculars and private affluence. But it was not to last. By the end of the fifteenth century Bruges was in competition with cloth manufacturing in England and Italy, and the city was losing its pre-eminent position. In an attempt to stop the Hanse merchants moving to another city, Bruges built them the Hanseatic House in 1483–7. It was an early form of prestige building, but it was too late. Cloth manufacturing and trade shifted its center of gravity and the city began a long decline. Merchants left, the canal silted up and Bruges never recovered. Its lack of subsequent development has meant many of the early buildings remain intact, so today Bruges is a tourist center where the discerning eye can see the built form of a typical medieval merchant city.

By the early 1400s Venice was the major city in the Mediterranean trading network. The city grew as its trade routes were underwritten by military and political power. Its informal empire was reinforced and protected by the presence of forts and a powerful navy. Venice not only traded goods, including cloth, silk, and spices, but was truly a merchant city, where an elite merchant class ruled for over 500 years. Commercial relations were regulated, trading rules were established, deals struck and contracts signed. Shared interests mingled with the pursuit of private gain. Trade, especially long-distance trade, was risky and merchants pooled their resources in shared enterprises. Because risk was spread through such collective endeavors a commercial civic culture was established. The ruling class maintained civic order by incorporating the prosperous middle class and the artisans in city deliberations and rituals, while still keeping a firm control over the principal levers of power. The main sites of the city were the business center of the Rialto and the political/public space of the Piazza San Marco. These two public spaces were linked by the Grand Canal, along the banks of which the most affluent merchants lived. They used the lower floors and attic for storage of goods and lived in the intermediate floors. As more wealth flowed into the city, the residences became palaces more than houses. The city was cosmopolitan, with substantial communities of Slavs, Greeks and Germans. Wealth from successful trading ventures also paid for art. Key figures in the Western art tradition, including Bellini (1431–1516), Canaletto (1697–1768), Titian (1488–1576) and Tintoretto (1518–76), received commissions from Venetian merchants. The city's wealth was embodied in an art of experimentation and innovation as pigments previously used only in manuscripts were painted on canvas. Venetian Renaissance art became known for it luscious, rich colors and its new subjects, including pastoral landscapes, dramatic scenes and portraits (Brown and Ferino-Pagden, 2006).

But by 1600 Venice had begun to decline as global overseas expansion favored cities on the western seaboard of Europe that enjoyed easier sea access to the

Atlantic and Pacific oceans. European merchants were extending their horizons beyond the Mediterranean basin to the wider world.

The early seventeenth century marked a transition in European and world affairs that involved a shift in global power northward, away from Italy and Spain toward France, England and the Netherlands. The first global power was Spain, its pre-eminence based upon its global empire in the New World and the Pacific. The sixteenth century belonged to Spain: it carved out a huge overseas empire in the Americas and amassed a considerable fortune, which allowed the Spanish Crown to pursue its role as defender of the Catholic faith in Europe. But the very success of its imperial position was to be the root of its subsequent decline. Spain illustrates an early example of what Paul Kennedy (1987) calls "imperial over-stretch." Superpowers arise on the basis of their military and economic strength. Challenges, however, are made to their dominance, and to maintain their position more of the resources are devoted to shoring up their geopolitical position and defending insecure frontiers. Imperial commitments therefore undercut the economic strength of the state. Spain was committed to ensuring its continued dominance in the Americas, safeguarding its commercial trade, and fighting numerous wars in Europe, especially against the Protestant English and the Muslim Turks. Imperial overstretch was heightened by the massive increase in the costs of war, the failure of the Spanish government to raise enough taxes, and the existence for Spain of "too many enemies to fight, too many frontiers to defend" (Kennedy, 1987: 48). There were recurring fiscal crises caused by increasing royal expenditure yet declining income as the mines in Spain's American empire began to be exhausted. By 1600 interest payments totaled two-thirds of all state revenues, and by the 1650s Spain was no longer a superpower.

If Spain was on a downward slide at the beginning of the seventeenth century, the Netherlands was on an upswing. It achieved wealth and power through the creation of a worldwide trading system. Trade and commerce were the lifeblood of the new republic. By 1600 the northern provinces had almost 10,000 ships sailing around the coasts of Europe and across the oceans. Trade was conducted in grain, tobacco, barley, herring, timber, sugar and spices. Dutch trading connections stretched around the Baltic and Mediterranean, across the Atlantic, south to Africa and across the Indian Ocean to India and Southeast Asia. At home a vigorous merchant community was successful in establishing a commercial society. The East India Company was established in 1602, and the Bank of Amsterdam in 1609. Soon both achieved prominent world positions. At this time, there was a wide variety of coins and currency. The Bank of Amsterdam took them all, assessed the gold and silver content, and allowed depositors to withdraw the equivalent in gold florins minted by itself. By doing this it became a depository of huge holdings and a central exchange of global financing.

Amsterdam was the center of the Dutch trading system. The city grew from only 20,000 in 1550 to almost 200,000 by 1700. Goods were brought from the rest of Europe, the New World and the Far East. Sugar refining, diamond cutting and tobacco curing all developed on the basis of imported goods. A mercantile culture was established. The art of the period reflected middle-class affluence rather than kingly splendor. The biggest building in the city was not a prince's palace but the Town Hall. Cosmopolitanism and religious tolerance became hallmarks of a city founded on trade and exchange. The newfound wealth was filtered through the moral membrane of a Calvinist theology. The dilemma of how to be wealthy and moral at the same time (Simon Schama (1987) calls this the "embarrassment of riches") gave shape and substance to a distinctly Dutch culture.

The Dutch Republic, and Amsterdam in particular, became a shipping center, commodity market and capital market for the world economy. In his 1728 *A Plan for English Commerce*, Daniel Defoe (1928: 123) summed it up thus: "The Dutch must be understood as they really are, the Middle Persons in Trade, the Factors and Brokers of Europe . . . they buy to sell again, take in to send out, and the greatest part of their vast Commerce consists in being supply'd from All Parts of the World, that they may supply All the World again."

Merchant capitalism, colonialism and global urban networks

To ensure a monopoly over trade, the Dutch, like their European competitors, needed to have control over local commerce. The prevailing ideology of the time was mercantilism, a belief that foreign trade was the chief method of increasing national wealth. The mercantilists believed that the world's wealth was like a giant cake, fixed in size: any increase for one nation was at the expense of the others. The most favorable conditions of trade were thus monopoly control. Successful commerce implied a commercial empire where prices could be fixed, markets protected and competitors kept out. The overseas expansion of European countries in the seventeenth and eighteenth centuries was a commercial undertaking driven by mercantile capitalism.

It was the need to ensure monopoly that led to the creation and maintenance of a global urban network of cities that linked Europe and the rest of the world. New towns were established and existing ones appropriated as command and control centers along the sea coasts of the world. Consider this account of the Dutch in North America that draws heavily on Short (2001).

The growth of the commercial empire involved the search for trading areas and territory in Africa, Asia and the New World. There was intense rivalry between the European powers for overseas trade and territory. North America was just one more commercial opportunity on the surface of the globe: most of South and Central America was under the control of the Spanish and Portuguese; in the far north France had already staked substantial claims. But between New Spain in the south and New France in the north the other European countries, including the Dutch Republic, sought trade openings and commercial opportunities.

The big prize for the Dutch merchants was Asia and the East Indies, which contained spices, timber, jewels and other exotic items that enabled merchants to make big profits. Trading companies were formed to pool resources and share the risk. The Dutch East India Company became hugely successful, paying annual dividends of never less than 20 percent and often as much as 50 percent.

In 1619 the Calvinists took control of the republic and were intent on a more aggressive policy of trade war against the Spaniards in the New World. The merchants of Amsterdam were also keen to open up new commercial possibilities in the New World. To secure commercial trade interests, the government of the Netherlands in 1620 established the Dutch West India Company. In this charter, which took effect in 1621, all trade from the Cape of Good Hope all the way west to New Guinea was in the hands of the company. Private traders caught breaking this monopoly were to have their ships and cargo impounded. The Dutch West India Company had attracted over 7 million guilders of investment by September 1623, a huge amount of money. It was given immense power, including the right to wage war, sign treaties, construct fortifications and encourage settlement to increase commerce and profit. It was divided into five chambers established in different towns, but the Chamber of Amsterdam was the wealthiest and took a special interest in the North American territory, called New Netherlands.

Colonization of free people was not the goal of the Dutch West India Company; first and foremost it was interested in maximizing profits. And in the New Netherlands they came primarily from the fur trade. (It was no accident that the seal of the New Netherlands depicted a beaver.) But the trade had to be established, controlled and monitored. To that end the company sent out thirty families in 1624. They established trading posts in its new territory: a commercial station on Governor's Island, just off Manhattan, and a trading post further upstream on the west bank of the Hudson, which they called Fort Orange.

The Native American Mohicans and Mohawks and the Dutch shared the same space, and interaction was common. There was the hidden exchange of viruses as the people of the New World succumbed to the diseases of the Old. Indigenous populations soon declined. A more visible Dutch–Native American connection was in the trade of furs. The Native Americans traded pelts for knives, muskets, drink, and tools such as hooks, kettles, hoes and axes. The trade changed Indian society: the tools lightened the domestic burden, the woolen goods led to changed

Case study 2.1

Madras/Chennai

European traders sought to control the trade with Asia, especially in the luxury goods of spice, cotton and silk. By the 1620s Europe imported 5 million pounds of pepper, 1 million pounds of cloves and nutmeg (combined), 350,000 pounds of indigo and 500,000 pounds of silk. Trade with Asia was a risky business. The sea routes were long and dangerous, the mortality rates on merchant ships was around 70 percent, items were expensive, so capital was tied up for long periods, and overheads were high. Permanent trading stations, called factories, had to be staffed and maintained. So trade with Asia was only possible by national governments or large trading companies. The English East India Company was formed in 1599 and a royal charter granting monopoly trading rights was issued in 1600.

The company set up trading settlements throughout Asia, including Madras on the southeast coast of India, where it built a fort and trading post in 1639–40. Soldiers garrisoned the site. Local merchants and cotton weavers were encouraged to set up business near the fort. By the beginning of the eighteenth century anywhere between 50 ships and 200 ships plied the important textile trade. Madras acted as a manufacturing center for silk and cotton textiles that were shipped back to London for the domestic market and re-exported from London to the rest of Europe. It was also a center for inter-Asian trade as private traders and merchants were involved in shipping textiles to the Persian Gulf, China and the Philippines. By the eighteenth century both English and Indian merchants were involved in this private trade. As with all trading cities there was competition. Madras's main rival was Calcutta, and there was conflict with the Dutch and the Portuguese for the lucrative spice trade and textile trade (Plate 2.2). The eventual British imperial dominance ensured the commercial success of the city in both the region and the wider world.

In the nineteenth century another economic layer was added to its functions, as it became an important administrative center. After Indian independence in 1947 it became a state capital. Founded as a commercial center, the city has managed to maintain its economic buoyancy. Textile manufacturing has been augmented by motor-vehicle and auto-parts production. Almost one out of every three cars built in India is made in Madras. The city is also a center for software development with much recent foreign investment. Its name was changed in 1996 from Madras to Chennai.

The city's trajectory from merchant company town to today's global metropolis is an excellent case study of urban economic growth and development in a colonial city. Its economic history is the story of changing connections with the global economy: from colonial outpost to hub in a new global network marked by global shifts in manufacturing, offshore investment and global outsourcing.

Plate 2.2 Portuguese colonial legacy in Chennai, India

clothing habits, while the liquor had a devastating effect. The furs were sent to Europe by the Dutch to be manufactured into hats, coats and other apparel. In 1624, 4,700 beaver and otter skins were sent to Holland, fetching 27,000 guilders. In 1635, 16,304 skins fetched 134,925 guilders. The fur trade was so lucrative that overhunting soon exhausted the supply, first in the seaboard areas and then in the Adirondacks. A distinction emerged between tribes who controlled fur resources and the seaboard tribes who no longer had fur to supply: the latter became more expendable. Obviously, in such a valuable business, bargaining was hard; the Native Americans wanted lots of goods for their fur, while the Dutch wanted to keep down the cost of the pelts. On both sides monopoly control was prized.

Good relations with the tribes were vital to the Dutch. The Native Americans provided food and controlled the supply of peltry, as well as land and information.

Officials were told "by small presents seek to draw the Indians into our service, in order to learn from them the secrets of that region and the condition of the interior" (Donck, 1968: 34). The greater the control over any or all of these resources, the more they figured in Dutch calculations. The West India Company produced a series of rules and regulations governing relations with the indigenous people, and officials were authorized to punish any Dutch who wronged a Native American. The emphasis was on orderly relations. The company was not in the business of saving souls. It wanted profit, and good relations opened the path to higher returns. Friction did occur, however, and the sharing of the same space was rarely harmonious. Dutch livestock trampled native corn, native dogs harassed settlers, there were arguments over women, and trading deals could break down into acrimonious dispute. There were arguments, fights and even all-out war. One director of the company, Willem Kieft, exacerbated tension when he became Governor by imposing a tax on the tribes around Manhattan in 1639. Passions were aroused, and from 1640 until 1645 there was sporadic fighting; almost a thousand Native Americans were killed. The Dutch lost people and property, and immigration from Holland, slight even in the best years, declined further. Kieft was replaced by Peter Stuyvesant and a treaty was signed in New Amsterdam in April 1645 in which Native Americans promised not to approach houses in Manhattan while armed and the Dutch pledged not to go near native settlements without warning while armed.

The Dutch settlement of New Amsterdam on the tip of Manhattan came under English control in 1664 and the town was renamed New York. It was one of a number of colonial implants around the world. At the same time that New Amsterdam was established, the Dutch had also founded Cape Town in South Africa and Batavia (now Jakarta in Indonesia). Other European powers were creating similar settlements. Not all prospered and not all grew into larger cities, but many did. And, of course, the three Dutch settlements all grew into major cities.

The European merchant city had two important urban legacies. The first was the creation of an urban culture in the major European cities, where there was the development of civic art, public spaces and collective rituals. These cities were the birthplaces of modern municipal government, including the creation of permanent city officials, the generation and spending of public revenues and the regulation of commercial affairs. Civic purpose was intertwined and often undermined by class, family and religious affiliations, yet private loyalties were tempered in the urban community of shared commercial interests. The European merchant city provides the template, model and background to the modern civic world and contemporary urban public culture. The second dimension was the creation of merchant cities around the world as command and control sites for

European merchant capitalism. The present-day global urban network owes its basic spatial structure to the merchant capitalism of European colonialism powers.

Cartographic representations of the merchant city

We have a visual record of the early mercantile cities (Short, 2004b). One of the largest (1.35 × 2.82 metres) and earliest (1500) printed bird's-eye views is Jacopo de'Barbari's illustration of Venice. The city is depicted, as if looking from the southwest, in a wealth of detail. It is a studio fabrication assembled from many drawings made in different parts of the city over the course of three years. Many other cities were also represented: Francesco Rosselli produced prospects of Pisa, Rome, Constantinople and Florence in the 1490s; an anonymous woodcut prospect of Antwerp appeared in 1515; a bird's-eye view of Augsburg was made in 1521 by Jorg Seld; and a woodcut prospect of Amsterdam by Cornelis Antoniszoon in 1544. Hans Lautensack produced a view of Nuremberg in 1552. Detailed urban maps celebrated the cities and had a talismanic quality: they were used to invoke good fortune and prosperity. They were also sources of civic pride, often used to illustrate local chronicles and proclaim the identity and prestige of a city. The level of detail in some of the illustrations also suggests the urban map as panopticon, a form of cartographic surveillance.

By the last third of the sixteenth century there was a considerable stock of urban maps and images. They had been drawn for a variety of reasons: civic pride; celebrations of specific events, such as the colossal prospect of Cologne by Anton Woensam drawn in 1531 on the election of Ferdinand of Austria as King of the Romans; military surveillance; parts of national inventories. Compilations of city maps and prospects were published in 1551 and 1567, but the first city atlas was the *Civitates Orbis Terrarum*. One volume was published in 1572, but it became so popular that by 1617 the work consisted of six volumes with over 363 urban views (Figure 2.2). Forty-six editions were produced in Latin, German and French. The atlas was so successful that it started a fashion that was to last into the eighteenth century.

The first volume of the *Civitates* was published in Cologne, edited by Georg Braun and engraved by Frans Hogenberg. It contains prospects, bird's-eye views and plans of cities from all over the world and provides us with a comprehensive collection of sixteenth-century urban views. In some cities individual buildings are named, while every city has a brief written note of its history, situation and commerce. The prospect and the bird's-eye view predominate, and even when the city is shown as a plan buildings are shown in vertical relief. The images also show the grandeur, wealth and power of the city. Each one is not just represented

Figure 2.2 Seville in *Civitates Orbis Terrarum* (courtesy of Library of Congress Map Collection)

but celebrated. Many of the urban maps and views were made to evoke and represent civic pride. The maps often adorned civic offices. In the atlas, some of the cities are shown with the loving detail of individual street names, buildings and churches. In many of the images, the cities come alive, animated by height and dimensionality; clearly a complex form of representation meant to honor them. The atlas rejoices in the urban condition.

Collectively, the images provide a comprehensive view of urban life in the Renaissance. They also indicate a world economy tied together in trade and linkages between urban centers. Aden, Peking, Cuzco, Goa, Mombassa, Tangiers and many other cities around the world are represented. The global reach of mercantile capitalism and European colonization is evident in the range of cities. While the cities are depicted separately, the effect of the compilation is to reveal a global economy of urban nodes and a trading world of connected cities.

Further reading

Braudel, Fernand, 1981–84, *Civilization and Capitalism 15th–18th Century*, London: William Collins' Sons. This three-volume work is a compendium of insights and empirical material on a crucial period in world economic development. Volume II, *The Wheels of Commerce* (1982), examines the social importance of the development of markets and merchant cities.

Hancock, David, 1995, *Citizens of the World: London Merchants and the Integration of the British Atlantic Community, 1735–1785*, Cambridge: Cambridge University Press. Case studies of merchant communities provide insights into the connections between place, people and commerce. This book is an invaluable case study as it gives insights into the lives, business practices and ideologies of a community of merchants at a particular time and a particular place. It examines the business and social strategies of twenty-three London merchants who developed the British Empire in the eighteenth century.

Jardine, Lisa, 1996, *Worldly Goods: A New History of the Renaissance*, London: Macmillan. Jardine questions traditional assumptions about the Renaissance by foregrounding the role of trade, commerce, material goods and conspicuous consumption. Her book shows the connections between commerce and art.

Matson, Cathy, 1997, *Merchants and Empire: Trading in Colonial New York*, Baltimore: Johns Hopkins University Press. Matson gives us a picture of the development of a merchant community in colonial New York from 1620 to 1770. She focuses on the middling level of merchants who resisted authority and developed an alternative set of ideas to mercantilism.

3 The rise and fall of industrial cities

Learning objectives

- To understand the links between industrial growth and urban development
- To examine the growth of the industrial city
- To explore the connections between the global shift in manufacturing and patterns of urban growth and decline

In the last chapter we looked at the rise of the merchant city. From 1400 to 1800 merchant cities were on the leading edge of economic and social change. The cultural achievements of cities such as Florence were built upon a vibrant trading economy and a dynamic banking center. Urban growth, in both its quantitative and qualitative terms, was built on trade and finance. Manufacturing played a part, as in the case of the Florentine cloth industry, but production was dominated by small-scale craft workers and limited production. A new engine of urban economic growth, industrial capitalism, developed around 1800 and it was embodied in a new urban form, the industrial city.

The Industrial Revolution was the result of complex set of factors that took centuries to develop; the notion of sharp break between a pre-modern, pre-industrial world and a modern, industrial world is therefore now discredited. For instance, while the Industrial Revolution developed in a series of technological breakthroughs built on measurement and calculation, the roots of this science lay as far back as the classical world, with the knowledge kept alive by Arabs and revived in late medieval Europe. Standardized production, as in the case of printing, was developed as early as the fifteenth century. The Industrial Revolution may have occurred around 1800, but the forces behind it had been fermenting for centuries.

At its core, the Industrial Revolution was new ways of making things. Production capacities were increased by mechanization. Steam power released capacities beyond the limits of human sweat. For instance, textile production was increased tenfold by the utilization of steam power. Once started, the Industrial Revolution expanded exponentially as inventions improved and streamlined the making of a huge variety of things. In the first wave emphasis was on textiles, but by the middle of the nineteenth century a second revolution centered on iron and steel production.

The Industrial Revolution has been theorized as part of long waves of production based on the clustering of innovations. The Soviet economist Kondratieff first identified these long, fifty-year cycles in the 1920s, and they are outlined in Table 3.1. Each wave is associated with key innovations that structure society and space. The first two waves, from 1785 to 1895, are associated with the development of factories and the growth of towns and cities. By concentrating production the Industrial Revolution promoted dense urban growth.

The first wave, from 1787 to 1845, is associated particularly with the UK. Britain was the first industrial nation for a number of reasons. It was a relatively small, densely populated country with relative political stability. Colonial expansion assured cheap imports and secure export markets. It is also important to note the role of the City of London. In 1600 it had a population of 200,000 and by 1800 almost 1 million, almost double the size of any other European city. Wrigley (1967) has argued that the city's population provided a huge, growing and secure

Table 3.1 The four Kondratieff cycles

	First 1787–1845	Second 1846–1895	Third 1896–1947	Fourth 1948–2000
Key innovation	power loom puddling	Bessemer steel, steamship	electric light, automobile	transistor, computer
Key industry	cotton, iron	steel	cars, chemicals	electronics
Industrial organization	small factories	large factories	giant factories	large and small factories
Labor	machine minders	craft labor	deskilled	bipolar
Geography	towns	towns	conurbations	new industrial regions

Sources: after Hall and Preston (1988); Short (1996: 73)

market for food producers and manufacturers, and this demand was the basis for further investment in agricultural and industrial production. To meet the demand, better roads were built, credit schemes were introduced and commercial relations were strengthened and deepened. There was therefore an urban bias to Britain's agricultural transformation and Industrial Revolution.

If Britain was the first industrial nation, it also had the first truly industrial city, Manchester. The English Romantic poet Robert Southey (1774–1843) described the city in 1807:

> The houses all built of brick and blackened with smoke; frequent buildings among them as large as convents, without their antiquity, without their beauty, without their holiness, where you hear from within, as you pass along, the everlasting din of machinery; and where when the bell rings it is to call wretches to their work instead of their prayers . . . Imagine this, and you have the materials for a picture of Manchester.
>
> (Southey, 1951: 23)

For the pantheistic Southey, the city represented a profane, godless place where work and machinery dominated, given over to Mammon. The new cities of the industrial age were a shock to the sensibilities. They were loud, noisy and dirty with the whole function of the city restructured towards making things, and especially profit. The Romantic Movement was, in part, a spirited response to this.

In 1800 Manchester was just one of hundreds of textile towns all over Europe. What made it the first industrial city were developments in technology, new sources of supply and demand, and social networks that fostered innovation and risk-taking. For centuries, textiles had been made by hand, but in the 1760s new steam-powered machines increased production. Previously, cotton had been imported from the Middle East and supplies were limited. But from the 1790s the US provided a vast, much cheaper supply. And the formal and informal British Empire provided a captive demand for manufactured goods. Cheap supply, new technology and steady demand all resulted in a rising industry. In 1774 the population of Manchester was only 41,032; by 1831 it was 270,901. And the mills were working day and night. Although it was based on textile factories, other industries also developed. To transport its goods, railways links were built: by 1840 the city was served by six railway lines and had become a center of locomotive construction.

The city was composed of three concentric rings. At the center was the exchange, surrounded by a warehouse ring, which was encircled by mills. Living conditions in the city were poor: industrial pollution and congested housing fostered high mortality rates. The life expectancy in the city was half that of the surrounding rural areas, for all socio-economic classes: in 1842 the average life expectancy of a laborer in Manchester was only seventeen.

Peter Hall (1998) asks the question: why did Manchester become the first industrial city? He focuses on the culture of innovation: between 1600 and 1800 a proto-industrialization system of economic organization was established; there was a capacity for continuous innovation; there was a large middle class of small

Case study 3.1
Locational theories of industrial cities

The rise of and fall of industry has been recorded and studied in a number of academic books and papers. Theories of industrial location explain the sites of economic activity. One of the earliest studies was Alfred Marshall's *Principle of Economics* (1891). A strong German contribution is embodied in Albert Weber's *Theory of the Location of Industries* (1909) and August Losch's *The Economics of Location* (1939). By the middle of the twentieth a body of work sought to explain the location of industry by looking at the various factors of production from the standpoint of the individual producing enterprise. Hoover's textbook (1948) is a summary work. Here is how labor costs are treated: "Low labor costs, an important locational factor for many industries, are found at several distinct types of location. The strengthened bargaining power of skilled and specialized labor in mature industrial centers has often encouraged a search for new locations and new processes adapted to these locations" (Hoover, 1948: 115).

A later work by Estall and Buchanan (1961) discusses the role of various factors, including materials, markets and transfer costs, energy sources, labor and capital, technological change and the role of government. In their case study of iron and steel, for example, they point to the pull effects of coal, iron ore, material assembly, market, capital and labor as well as to the importance of inertia.

Critics of standard industrial location theory pointed to the assumption of perfect economic rationality. Decision-makers, in the models at least, had all the necessary information and were blessed with perfect rationality. A number of scholars pointed to the need to incorporate suboptimal decision-making and satisfying rather than maximizing behavior. A more behavioral approach is summarized in Hurst (1972). Critics also pointed to the emphasis on single-plant, single-product establishments and raised the need for a fuller understanding of agglomerations, corporate context and multinational transactions. An institutional approach emerged that stressed the interaction between firms rather than the behaviors of individual firms, and showed that location was the result of complex negotiation and bargaining with a variety of other agents in a wider social and political context (Martin, 2000). More recently, some economic geographers have utilized evolutionary economics to refine location and relocation theories (Boschma and Frenken, 2006; Cooke and Morgan, 1998).

capitalists able to employ a range of entrepreneurial talent, social networks and local cultures that allowed a constant improvement in products and processes. The city's location was also important: it was close to coal that provided the raw power that animated the incessant spinning and weaving machines. But other places were close to coalfields too, so Hall (1998: 347) argues that Manchester became the first industrial city primarily because it was "the first true innovative milieu."

The industrial city and social conflict

While the city was a center of innovation, it was also a scene of urban conflict. One of the earliest commentators on social conflict in the industrial city was Friedrich Engels (1820–95), who had more than a theoretical interest in capitalism: he was an active participant in one of the leading sectors of the era, the textile trade. He worked for an export business in Bremen as a young man and moved to Manchester in 1842 to work in a branch of the family business as a manager in the Cotton Exchange. His experiences over the next two years in Manchester prompted him to write *The Condition of the Working Class in England*. Engels describes an Industrial Revolution driven by technological changes such as the invention of the spinning jenny in 1764. The rise of the power loom, mechanical power, iron smelting and railroads all reinforced the concentration of economic activity in towns. He describes the working conditions of working people and documents their dreadful living conditions, their high death rates, their inadequate food and their poor housing.

Engels met Karl Marx in August 1844 in Paris and they began a lifelong collaboration. For both men, the new industrial city was intimately connected to the new capitalist mode of production: it embodied all the paradoxes and revolutionary potential of capitalism. In *The Communist Manifesto*, written in 1844 and first printed in English in 1888, they noted,

> "The bourgeoisie has subjected the country to the rule of the towns. It has created enormous cities, has greatly increased the urban population as compared with the rural, and has thus rescued a considerable part of the population from the idiocy of rural life".
>
> (Engels, 2004: 5)

So, for optimistic radicals with a belief in the forward march of history, the city saved people from the impediments of rural life by giving them experience of their collective strength. Marx made a distinction between classes in themselves and for themselves. While classes could be objective facts (classes in themselves), in order to become agents of history they also needed to become aware of themselves as products of history with an ability to make a new and better future. Engels and Marx were writing at a time when a revolutionary capitalism was sweeping away much of the pre-modern world. "All that is solid melts into air" is the famous phrase that captures their sense of this historical rupture. The industrial city, as the cauldron of the newly emerging working class's concentrated power, was the main fracture point.

The basic Marxist theory declares that rapid industrialization and urbanization lead to class formation and class identity. The testing of this idea is the basis of a major work of historical scholarship, E. P. Thompson's *The Making of the English Working Class* (1963). Class and class relations are pivotal to his analysis of the roles of urbanization and industrialization in the formation of the working class. The title of the book implies a double making of "working class" as Thompson examines how it became a class *in* itself and *for* itself. A product of rapid urban industrialization, the English working class also made itself through its sports clubs, burial clubs, political affiliations, religion, civic socialism and trade unions. The working class that Thompson identifies demanded and secured substantial gains from what was originally a very oppressive system. Rather than a revolutionary break, Thompson reveals the working-class dynamic more as a steady drive toward the improvement of working and living conditions within an existing system.

Subsequent work has tempered the early Marxist notion of the workers of industrial cities as gravediggers of the capitalist order. John Foster (1974) examines working-class attitudes in three English cities, focusing particularly on Oldham. He shows that from the 1790s to the 1830s a labor consciousness was formed in the industrial cities that, by the middle of the nineteenth century, had crystallized into a revolutionary working-class consciousness. This was just when Marx and Engels were casting their critical eyes over class relations in England. However, by the end of the century, labor had been subordinated to the capitalist order. Foster points to the active attempt by the bourgeoisie to win back mass allegiance. By the 1860s, one-third of all engineering workers and one-third of all male workers in cotton manufacturing were exercising authority over fellow-workers. In other words, the working class had become more segmented, with substantial portions more aligned to management in the new divisions of labor. Foster's work shows that revolutionary class consciousness was a temporary phenomenon before the capitalist class won back allegiance and before changes in labor organization resulted in new labor aristocracies.

Patrick Joyce (1980) looks at the social consequences of mechanization and community development in factory towns of the later nineteenth century. The factory became a place where work got under the "skin of life," a place that promoted both a defensive form of class solidarity and a culture of subordination.

However, while revolutionary class consciousness may have been tempered, the organized working class provided a new, important social power base in industrializing economies. Together with its political representatives it success-fully bargained for a higher level of social welfare that in Western Europe resulted in easier access to existing public goods, such as healthcare, and the creation of "new" public goods, such as social housing. Indeed, the political power of the organized working class in Western Europe managed to transform many goods from private (with access based on income) to public (with access based on citizenship).

As industrial capitalism matured, the factory and the industrial city did not produce the shock troops of revolution. Rather, industrial workers became a source of social stability. Later studies of twentieth-century factories, such as Benyon's (1973) study of car workers in a British Ford plant and the Nichols and Benyon (1977) analysis of workers in a chemicals factory, show that organized labor was more concerned with striking shop-floor bargains than with broader political objectives. There were also national differences in the political role played by organized labor. Scandinavian countries have a much more comprehensive welfare system than the US. Katznelson (1979) shows that ethnic cleavages, reinforced by strong residential segregation, was one reason behind the lack of class-conscious labor politics in the US, where unions tended to focus on gaining factory- or at most industry-wide concessions rather than broad social welfare provision. Race trumped class in most US cities, in part as a result of urban segregation which in some cases was an explicit urban policy. Charles Connerly's (2005) study of the steel town of Birmingham, Alabama, reveals that city planning until the 1950s was dominated by the desire to segregate the races and to ensure that African-Americans lived in isolated, poorly maintained communities.

The planned city

The rapid growth of town and cities, especially in the wake of rapid indus-trialization, created a series of urban problems: inadequate housing, polluted environments and poor public health. The responses to these problems shaped the nature of the planned city. Three responses can be noted: social activism; new models for cities; and urban interventions. Let us consider examples of each.

Many social activists were motivated by the abysmal conditions of the indus-trializing city. In 1889 Jane Addams founded Hull House on Chicago's Near West Side. She worked there until her death in 1935. Addams and the other residents of the settlement provided services for the neighborhood, such as kindergarten and daycare facilities for children of working mothers, an employment bureau, an art gallery, libraries, and music and art classes. Hull House surveys of the local area created maps of household income levels and ethnicity. Eight out of ten of the contributors to 1895's *Hull House Maps and Papers* were women. By 1900 Hull House activities had broadened to include a co-operative residence for working women and a meeting place for trade union groups. The Hull House residents and their supporters forged a powerful reform movement that launched the Immigrants' Protective League, the Juvenile Protective Association, the first juvenile court in the nation and a Juvenile Psychopathic Clinic. They lobbied the Illinois legislature to enact protective legislation for women and children and to pass in 1903 a strong child labor law and an accompanying compulsory education law. The federal child labor law of 1916 was the national result of their efforts.

Ebenezer Howard's ideas of new, planned, garden cities were outlined in his 1898 book *To-morrow: A Peaceful Path to Real Reform*. It was reissued four year later with a new title, *Garden Cities of To-morrow*. Howard had a vision of small co-operatives, self-governing settlements. He could see the ills of the big city and the unemployment in the rural areas and suggested a garden city in the countryside where land could be bought cheaply and decent housing made available. It was not a rural retreat: Howard expected that jobs and industry would also leave the big city. The ideal settlement size Howard proposed was 32,000 people at relatively high densities in a city surrounded by a green belt. Once urban growth exceeded this limit, a new settlement would be established and eventually a vast conurbation would emerge of multiple garden cities linked by a mass transit system. Residents would own the land and modest rent levels would be used to fund public services and social welfare programs. Howard combined the idea of garden cities with communal land ownership and local self-government all set within a vast urban region. Garden cities were built around this time. Letchworth, just north of London, was the first, built in 1903 with Raymond Unwin and Barry Parker as the main planners. Unwin and Parker went on to build more garden cities, garden villages and garden suburbs. Thirty-two new towns were built in post-war Britain and at their height they housed almost 2 million people. The idea also diffused out from Britain. Radburn, in New Jersey, was the first garden city in the US, begun in 1929.

However, while new towns were built the co-operative land schemes were rarely implemented. The design elements of garden cities and garden suburbs became staples of urban design, but not the physical blueprint for a new social order. The most recent garden cities are private developments, often gated communities built by private interests for private gain.

Other urban designers had visions of new cities but without radical social alternatives or commitments to social reform. These more design-minded urban visionaries included Baron Haussmann, Daniel Burnham and Le Corbusier. The modern city bears the imprint of their visions.

Paris, like many other rapidly growing cities of the nineteenth century, had substandard housing, poor public health and a disenfranchised population that could be a threat to the existing political order. The old neighborhoods, especially in the east of the city, played an important role in the revolutions of 1830 and 1848. The Emperor of France, Napoleon III, charged Haussmann with making Paris a spectacularly beautiful city, a healthy city and a city where the mob was under firm control. Haussmann demolished the oldest neighborhoods in the medieval heart of the city, destroying the homes of 15,000 people. He built wide, straight boulevards that made it more difficult to put up barricades and easier to move troops (Plate 3.1) and effectively moved the center of Paris to the north west, considered a safer part of the city, where the rich lived. Haussmann radically altered the city in strokes that were repeated in many twentieth-century cities as urban renewal schemes demolished old neighborhoods.

Burnham drew upon Haussmann's designs in emphasizing ceremonial public spaces and long vistas ending in neoclassical buildings. He designed early skyscrapers and the Columbian Exhibition (held in Chicago in 1893), and was

Plate 3.1   Haussmann's Paris

one of the originators of the City Beautiful Movement that sought to rebuild public spaces with grand designs and imposing processionals. He designed a new civic center for Cleveland and in 1909 devised an ambitious plan for Chicago that imposed a classical civic order on the regular grid. By 1925 much of his plan had been completed. Peter Hall (1988) shows how the City Beautiful designs spread around the world.

The Fordist city

The first Kondratieff cycle saw its full flowering in Britain: Manchester was the exemplar nineteenth-century industrial city. The second and third waves were experienced in other parts of the developed world, Germany and the US in particular. By the time of the third Kondratieff cycle, based on automobile manufacturing, giant factories and standardized deskilled manufacturing processes, the epicenter had shifted decisively to the US. The exemplar city was Detroit as it became home to car manufacturing and new forms of industrial production. At the beginning of the twentieth century cars were luxury items, handcrafted and designed for the wealthy. Detroit was only one of many car-making cities. Even before the coming of car production it was a manufacturing center, and by 1900 it boasted almost a thousand machine shops making ships, stoves, engines and mining machinery. There was a pool of skilled labor and network of local financiers. It was soon producing two-fifths of the nation's car output, concentrating on the cheaper end of the market.

Henry Ford transformed the manufacture of cars. He was born in 1861 and grew up close to Detroit. In 1879 he went to work for the Michigan Car Company that was building ten cars a day. Twenty years later he formed the Detroit Automobile Company, but it was not a success: Ford was too much of a perfectionist and had yet to hone his market sensibilities. Fine-tune them he did, however, when he founded the Ford Motor Company in 1903 with twelve shareholders, all from Detroit. In 1908 the first Model T appeared on the market. It was a simple yet robust marque that went through numerous design improvements to become one of the first mass-produced cars. Assembly lines and mass production had been developed over the previous century in a range of manufacturing industries, including bicycles and watches as well as the Chicago meat-processing and packing industry. So Ford did not invent mass production, but he refined and improved it. Car production became standardized, precise and continuous. Mass production also allowed reduction in the final price: the cost of a Model T in 1908 was $805; by 1924 it was down to $290. With each price decrease new markets were created, so mass production not only met demand but created new demand. Cars became less a luxury item and more a regular purchase, especially in the rural hinterland where farmers used them for a variety of purposes. By 1920 every second car in the world was a Model T Ford. In order to reduce labor turnover Ford also paid relatively high wages. A well-paid industrial workforce was essential to decreasing labor turnover and maintaining worker allegiance. So Ford was instrumental in creating the high-paid blue-collar sector of the new industrial city.

Detroit was the forerunner of the modern industrial city. Production was controlled by a small number of very large companies who operated under oligopolistic conditions. Production was mechanized, repetitive with long production lines. An organized working class was both relatively well paid and well cared for. There were outbreaks of labor unrest when the business cycle softened demand and management implemented layoffs and wage cuts, but, by and large, a stable system of capital–labor relations was established, a relatively affluent working class was created (in the US it saw itself as blue-collar middle-class) and the power of well-organized labor was reflected in the developed countries with high social welfare provision and government commitments to full employment. The high point of what has been termed the "Keynesian–New Deal city" (Short, 2006a) lasted from 1945 to the mid-1970s. Thereafter, this particular form of the industrial city was transformed in three ways: manufacturing decline, postfordism and global shift.

Manufacturing decline

Manufacturing decline is a worldwide phenomenon: in 1995 there were 172 million manufacturing jobs, but by 2005 this had declined to 150 million. In 1995 China had 100 million factory jobs; since then between 10 million and 20 million have disappeared. Over the same decade in the UK the number of factory workers declined from 4.7 million to 3.6 million. In the West Midlands the number of manufacturing workers declined from 629,000 to 444,200, with the city of Birmingham losing 38,900 jobs from an initial total of almost 100,000. In South Korea companies are moving their plants to offshore sites with cheaper labor and easier access to new markets. Between 2005 and 2006 the country lost 75,000 manufacturing jobs. In countries, regions and individual cities around the world, the decline of manufacturing employment is an economic change of seismic proportions.

Consider the case of the US. Manufacturing employment grew from around 15 million in 1960 to its peak of over 19.5 million in 1979. Since then, while there has been some fluctuation, the general trend has been downward, with a marked decline since 2000. In January 2004 the number of manufacturing jobs stood at 14.3 million, with a loss of 3 million from 2000. In total, almost 6 million manufacturing jobs were lost from 1979 to 2004, and manufacturing as a share of the nation's economic output fell from 21 percent to 14 percent. Behind this change lay important social implications as a heavily unionized workforce in relatively high-paying, secure jobs expanded then contracted, and traditional industrial districts and cities lost jobs and economic rationale.

There are three main reasons for the decline in manufacturing employment in the US and other industrialized nations. First, there was a slower rate of growth in the demand for manufactured goods. As consumers' income rose and the economy matured, there was a shift from spending on goods to spending on services such as healthcare. In 1979, 52 percent of consumer spending was on goods; by 2000 it stood at only 42 percent. Second, there has been an enormous increase in manufacturing productivity. Each year since 1979 the productivity of manufacturing workers has increased by 3.3 percent, which compares favorably with the rest of the economy, where productivity growth rates are closer to 2 percent. As machines replace people, fewer workers are needed to produce the same amount of goods. The automation of production reduces the power and influence of skilled manufacturing workers. Third, there has been growing competition from overseas manufacturers. Goods formerly produced in the US are now being made more cheaply offshore, especially in countries where wage rates are substantially below US levels.

But behind this general decline was the growth of manufacturing employment in selected parts of the US. In sectors of Los Angeles and New York, for example, clothing manufacturing continues to employ many people, mainly immigrant women. And new car factories have sprung up in the small towns of the South and Midwest. While many older automobile manufacturing plants closed, new automobile production and assembly plants opened along a seventy-mile band running alongside Interstate Highways 65 and 75 from Michigan through Indiana and Ohio into Kentucky and Tennessee. Between 1980 and 1991, twenty new vehicle assembly plants were built in this corridor to minimize freight costs, and located in small towns to avoid heavily unionized areas.

The contraction of manufacturing employment varied. The largest declines were in the old established urban centers of the Northeast and Midwest, traditionally unionized places, while the largest relative and absolute increases were in the Sunbelt states, many of them with anti-union laws and regulations. The decline of manufacturing in the old established cities effectively weakened the hand of organized labor and strengthened the hand of capital. The result was a squeeze on the living standards of industrial workers and a general depression of middle-class incomes. The average weekly earnings of all private industry employees declined from $275 in 1982 to $274 in 2001 (in constant 1982 US dollars).

Deindustrialization involves the closure of plants, especially in the urban cores, the reduction of workers in existing plants and a net shift in manufacturing employment away from the central city to the suburbs. In vibrant urban economies the loss of manufacturing jobs is offset in aggregate terms by the growth of jobs in other sectors. However, in neighborhoods and cities where there is little alternative employment growth, the loss of manufacturing employment is devastating. Traditional small industrial cities such as Schenectady, Syracuse and Flint in the US saw the lifeblood of the local economy drained away without the transfusion of new jobs. In selected inner-city neighborhoods jobs have gone, not to be replaced. Dundalk, a blue-collar Baltimore suburb, saw a decline in manufacturing from 48 percent in 1970 to 16 percent in 2000. With few employment opportunities, the poverty rate doubled from 5 percent to 10 percent. William Julius Wilson (1996) charts the effects of such employment loss on minority neighborhoods in such cities as Chicago, where deindustrialization impacts significantly on African-American urban communities.

Detroit embodies the process of deindustrialization and job loss. The city grew in the first two-thirds of the twentieth century on the back of the car industry. In 1960 there were 642,704 jobs in the city and the population was 1.6 million. By 2000 there were only 345,424 jobs, the population had declined to just over a million, vast swaths of the city had become vacant and an endemic fiscal crisis limited the effectiveness of city government. Motor City became Vacant City as 17,000 acres out of the city's total of 86,000 lay vacant and derelict.

The Postfordist city

From the 1970s two major changes took place that constitute the maturing of the fourth Kondratieff cycle. In terms of internal organization increasing competition and a rise of consumer power prompted a change in industrial production from a narrow reliance on a few items to a more varied inventory designed to meet a fickle market (Webber and Rigby, 1996). There was also a global reorganization of manufacturing production, best described as "global shift" (Dicken, 2007).

Postfordist techniques involve greater flexibility of production, more subcontracting, greater vertical integrations and more flexible labor. Rigid craft distinctions and demarcated job specifications are loosened as management uses labor in a more flexible manner. The reorganization of production allows firms to adjust quickly to a more volatile market, permits higher rates of productivity and reduces employment costs. It also signals a weakening of the power of organized labor as management regains more control over the deployment and the pace of work. Postfordism (the terms "flexible specialization" and "flexible production" have also been used) therefore provides capital with more control over labor.

Flexible production is also linked to a spatial reorganization from the standardized space of factories as nodes of giant corporations toward more vernacular production spaces of smaller firms. By the end of the twentieth century the old industrial cities of Detroit were losing out to the new, looser agglomerations of Silicon Valley.

The shift to flexible production is not limited to the developed world. Christerson and Lever-Tracy (1997) point to the dense networks of small firms emerging in rural China that are competitive in rapidly changing global markets. Postfordist industrial districts are emerging in a select group of developing countries.

In summary, since the 1970s, manufacturing has shifted toward flexible specialization involving shorter production runs, more automation and less skilled workers. While this makes industry much more sensitive to changing consumer

Case study 3.2

Contrasting urban fortunes

In 1950 Schenectady, New York, and San Jose, California, had similar population sizes: 92,061 and 95,280, respectively. At that time Schenectady's economy seemed secured. Over 27,000 people worked for General Electric. The city was flourishing, with a vibrant downtown and a buoyant job market. Corporations such as the American Locomotive Company provided employment for thousands of workers as well as a way of life. Social clubs and softball teams grew up around the connections workers made in the factory. Fifty years later most of those jobs had gone. The Locomotive Company closed in 1968, and General Electric shed over 90 percent of its workforce in the city. The city's population shrank to just over 61,821; the houses lost value and the credit rating of the city, which always provides a significant fiscal view of a city's economic health, was downgraded by Moody's Investors Service to the lowest in the state. The median household income in 2004 was just $29,378.

By 2000 San Jose, in deep contrast, had a population of 894,943 and was one of the larger cities in the US. Decades of spectacular growth fueled in particular by the Silicon Valley boom in high-technology and computer-related industries make San Jose one of the most prosperous and economically dynamic cities in the country. In 2004 median household income was $70,243. Among the companies headquartered in the city are Adobe, Cisco and eBay.

Schenectady and San Jose: close in population size in 1950, but by 2000 they were on two very different trajectories, one spiraling downwards and the other surging upwards.

preferences, it has reduced the number and skill levels of manufacturing workers. This in turn has led to a decrease in labor's bargaining power with resultant effects on wages. And the more jobs are deskilled, the more easily they can be shipped offshore to unskilled labor pools.

In the shift from manufacturing to a service-dominated economy, one particular sector has emerged as a very important economic and cultural phenomenon: retail. This constitutes one of the single biggest employment sectors, larger than the manufacturing or government sectors and second only to the service sector. Retailing is therefore a central feature of economic organization. The old push economy of the first three-quarters of the twentieth century – in which companies made things and retailers then sold them to consumers – is now a pull economy, in which large retailers dominate the consumer market. The 'big box' retailers, such as Wal-Mart in the US, have so much power in the market place that they dictate to manufacturers the cost, style and delivery schedule of goods. The relentless drive toward reducing costs by such powerful companies as Wal-Mart is central to the staggering productivity increases in the manufacturing of consumer goods and to the globalization of manufactured goods. As Wal-Mart forces producers into intense competition to trim costs, it compels many US companies to turn to China in order to keep up with its Asian suppliers. The large retailers have so much power, they force manufacturers not only to reduce costs but to move production offshore. Wal-Mart's inventory is now dominated by cheap imports from Asia, and while this provides lower prices for US consumers, it also disciplines manufacturers and ultimately US workers. Wal-Mart provides a platform for Asian goods and lubricates the globalization of manufacturing toward lower-cost producers.

Capitalist economies are regulated. Markets exist in legal, social and cultural contexts, and operate in and through such contexts. There is no such thing, except in the minds of neoclassical economists, as a "pure" market. The term "post-fordist" has also been associated with changes in state policies as Fordism involved a regime of regulation that included collective bargaining between big labor and big capital. The state set the framework for this negotiation but also was involved in maintaining low unemployment levels. After the mid-1970s as flexible specialization and manufacturing decline weakened the hand of big labor, a new, more entrepreneurial state began to emerge that was more responsive to the needs of business. The state became concerned with wealth creation rather than wealth redistribution and with employment creation rather than employment protection. Deregulation, privatization and a reduction in social welfare became parts of a neo-liberal agenda that dominated state polices (Dumenil and Levy, 2004; Harvey, 2005).

Global shift

The economic geographer Peter Dicken (2007) uses the term "global shift" to describe the movement of industrial employment from the cities of the developed world to those of the developing world. There has been a redistribution of manufacturing employment from Western Europe and North America to Asia that entails the deindustrialization of the advanced economies and a rapid industrialization in selected cities in a small group of developing countries, including China, Singapore, Taiwan and South Korea.

Global shift is moving up the manufacturing chain. In the early 1990s China's economic growth was dominated by traditional, labor-intensive manufacturing sectors such as textiles, clothing and footwear. More recently, however, growth is more noticeable in capital-intensive, high-tech sectors, such as machinery and electronics. Industrial cities in countries like China that attracted labor-intensive industries are now also attracting capital-intensive industries (Wong and Chan, 2002).

We can focus on global shift by looking at just example, the case of Nike (Short, 2001). Phil Knight was a member of the University of Oregon track team in the 1950s. His coach was Bill Bowerman. Knight went on to Stanford Business School and, when faced with a term paper, he developed the idea that low-cost Japanese shoes could find a market niche in the US athletic shoe market. He did not pursue the idea immediately, becoming an accountant in Oregon, but on a trip to Japan in 1963 he picked up a pair of Tiger running shoes. He showed them to Bowerman, who thought they were better than Adidas shoes. They invested one thousand dollars in a thousand pairs of Tiger shoes and sold them at local high school track meets. It was the beginning of a lucrative connection. Bowerman would send new designs to Japan and new shoes would be made, shipped back to Oregon and sold at track events. By 1969 the annual sales were almost a million dollars. At this stage Knight and Bowerman were selling the shoes under the original Japanese brand names. However, in 1971 Knight decided it was time for a separate identity. The shoes were named after the Greek goddess of victory, Nike. The swoosh symbol was designed by a Portland design student in 1972. Annual sales that year were $3.2 million. By 1980 they were $270 million and one out of every three Americans owned a pair of Nikes. By 2000 Nike was selling close to 100 million shoes per year and generating annual revenue of $10 billion.

Making shoes is dirty, dangerous and difficult. Initially, Nike shoes were made in Japan. The Japanese perfected their designs and the products they sold on the world market continued to improve. So did Japanese workers' wages and conditions. Labor costs rose. In 1974 Phil Knight made his first visit to South

Korea. By the early 1980s most Nike shoes were made in Korea, and the city of Pusan became the capital of Asian shoe manufacturing. Nike signed contracts with Korean shoemakers. Factories sprouted up, more workers were employed. South Korea became one of the fastest-growing manufacturing nations in the world. Meanwhile, during the 1970s and 1980s 65,000 jobs in shoe manufacturing in the US were lost. By the mid-1990s a pair of Nikes that sold for $30 cost just $4.50 to make. Line workers in Korea were receiving $800 a month. But in the competitive shoe business further cost reductions were needed. Nike could reduce their costs by getting their shoes made in China, Indonesia and Vietnam, where labor costs were only $100 a month. In Vietnam in 1998 workers at a Nike shoe manufacturing plant earned as little as $1.60 a day. In Indonesia, workers were sometimes receiving as little as 50 cents a day. Indonesia is now one of the largest suppliers of Nike shoes: 17 factories employ 90,000 workers, producing around 7 million pairs of shoes each year. In southern China the center of shoe manu-facturing is the city of Guangzhou. Just outside the city one shoe factory used by Nike make 35,000 shoes every day.

The story of Nike is one that can be found in most manufacturing sectors. In the 1950s the bulk of the world's manufacturing jobs were in the old industrial heartlands of North America and Western Europe. This was the result of an international division of labor that had emerged over the previous 200 years. The core of the world economy imported raw materials, turned then into manufactured goods and exported them around the world. This division of labor involved an unequal exchange in which the core economies of Europe and North America grew richer. Cheap raw materials rarely provided the basis for industrial take-off in the periphery; but in the core economies, the value-added work of manu-facturing provided the basis for sustained capital accumulation. It was also the basis for the creation of an organized working class.

The global shift in manufacturing employment has meant a reconstruction and reterritorialization of the working class. A predominantly male, North American/ European working class, earning union rates, has been replaced by a young, female, Asian working class. The old-established, self-conscious working class has been effectively destroyed, while the new one has yet to organize itself sufficiently to exercise political and economic power. In the new global world, capital can move to cheaper labor areas; while organized labor in traditional manufacturing industries, such as textiles and shoe manufacturing, has seen the jobs move away. Economic globalization allows corporations to relocate in order to minimize wage costs. There has been a deterritorialization of corporations, so the old adage that what is good for General Motors is good for the US no longer applies. Although listed as a US company, Nike's interests do not necessarily parallel US interests. What is good for Nike's shareholders is not necessarily good for US workers. The low cost of international transport and the growing ease of international trade, crucial requirements of economic globalization, have allowed capital to be more easily dissociated from national interests and local community concerns. Capital is free to roam the world in search of ever-lower wages. Globalization has liberated capital from territory, citizens and communities. It now signals the power of capital to move at will, while those without capital are stuck in place. Space and place; freedom and constraint. Globalization gives more power to the powerful and further constrains the weak.

The mobility of capital has been reinforced by changes in production. In the Fordist model plants were huge, fixed-capital investments. Bargaining between capital and labor thus took place in a fixed location. More recently, a more flexible form of production has been introduced. Nike has no shoes factories and there are no Nike workers. The company's shoes are made under contract by a range of manufacturers. Factories compete to obtain Nike orders and are then licensed by Nike if they are capable of making shoes to cost and design specifications. Many of the "Nike" shoe factories in Indonesia and China are actually owned by Korean and Taiwanese business interests. This system drives down prices. The old model of manufacturers making things and retailers merely selling them has been replaced by the power of retailers and brands. Now retailers tell the manufacturers what to produce. Contracts are for short-run lines rather than long-run batches. One clothing retailer, Hennes and Mauritz, a Swedish company with stores in Europe and North America, keeps its prices low by contracting in low-wage areas of the world. Almost 900 factories are used to produce a constantly changing design portfolio. The company has been successful in keeping its inventory low; the just-in-time production system ensures that goods are made to meet demand, with stores often receiving daily supplies. The entire inventory is turned over eight times each year (the industry standard is four times). High turnover means that profits can be made through selling many items rather than one; hence the price of individual items can be reduced, which in turn aids turnover. Designs seen in Italian fashion shows are soon produced cheaply and quickly and then sold in stores until the next fashion wave hits. Just-in-time, flexible production allows low prices and high turnover, and is indicative of a marked change in capital–labor relations. Capital is now hypermobile. Workers in one factory cannot bargain in the same effective way that the workers of the old Ford system could. Capital is no longer fixed in place. Retailers can move their production contracts to another factory in another country. While capital can roam the world, labor is fixed in place. The result is an uneven bargaining arrangement.

One consequence of global shift has been an increase in international trade. An increasing share of spending on goods and services is devoted to imports from other countries. The share of international trade in total output rose from 27 to 39 percent between 1987 and 1997 for developed countries. The corresponding figures for developing countries were 10 and 17 percent. The production chains of international corporations such as Nike now weave their way through many countries, and the empirical evidence suggests that there is a correlation between international trade and per capita income. Less protectionism means competition, greater awareness of new foreign ideas and technologies; and, for poor countries, the ability to import capital equipment necessary for long-term economic growth. The result is more efficient economies. More open trade leads to rising per capita income. However, per capita income is a crude measure that does not show the distribution of income, simply the average. The evidence on income distribution is inconclusive. In several countries income inequality has *increased* in the wake of trade liberalization: Argentina, Chile, Colombia, Costa Rica and Uruguay, for example. In the US the wage rates of high-school educated males fell by 20 percent from 1975 to 1995 as the US economy became more open to foreign imports. It is difficult fully to assess the role of trade in this inequality because technological changes that weaken the position of selected labor groups must also be taken into consideration. However, the perception of rising inequality is apparent, and it is most often associated with foreign trade. In the US, for example, the declining incomes of the middle class are commonly associated with cheap imports, foreign workers, capital disinvestment, and reinvestment in cheaper producing areas and globalization in general.

Economic globalization is a work in progress rather than an achieved end point. Not all firms are footloose, and technology transfers between countries remain problematic. A truly global economy would have a free transfer of capital *and* labor. What we have is free movement of capital while labor is increasingly state regulated. Moreover, not all countries have been involved in the global economy. The so-called Asian "tigers," despite recent crises, have dominated the global shifts in production. Japan, Hong Kong, Singapore, South Korea, and Taiwan were the first wave of newly industrializing countries. More recently, China, Indonesia, the Philippines, and Vietnam have attracted the bulk of investment in manufacturing (Plate 3.2). Although we may think of global production chains, in practice only a few countries are involved: Nike's Air Max Penny shoes have 52 components from only five countries – the USA, Taiwan, South Korea, Indonesia and Japan. Economic globalization is spotty. Much of sub-Saharan Africa has been excluded from both the first and the second waves of new industrialization. A global economy is becoming rather than being, and it is selective, patchy and incomplete.

**Plate 3.2
Industrial park
in Ho Chi Minh
City, Vietnam**

Further reading

Dicken, Peter, 2007, *Global Shift: Mapping the Changing Contours of the World Economy*, New York: Guilford Press. Now in its fifth edition, this book highlights the creation of the global economy and its changing configurations. A must-read work.

Hall, Peter, 1998, *Cities in Civilization*, London: Weidenfeld Nicolson. A magisterial summary of cities as centers of innovation. While the book ranges across 2,500 years, the chapters on nineteenth- and twentieth-century cities, including Berlin, Detroit, Glasgow, London, Los Angeles, Manchester, New York and Tokyo, provide unrivaled and detailed empirical material on the industrial and post-industrial city.

Internet Modern History Sourcebook, 1997, "Industrial Revolution" (<http://www.fordham.edu/halsall/mod/modsbook14.html>). A useful website on the Industrial Revolution.

Webber, Michael J. and David L. Rigby, 1996, *The Golden Age Illusion: Rethinking Postwar Capitalism*, New York: Guilford. Provides a detailed discussion of the global industrial growth of the 1950s and 1960s and decline since the mid-1970s. An important discussion of the global spatial reorganization of capitalism.

4 Service industries and metropolitan economies

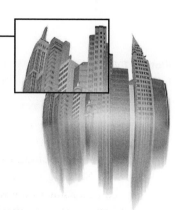

Learning objectives

- To understand the shift from manufacturing to service employment in urban economies
- To see the importance of advanced producer services
- To visualize a global urban network of service centers

While merchant cities such as Venice and Bruges reached dazzling degrees of civic splendor, and industrial cities such as Manchester and Pittsburgh grew into global prominence, they also all witnessed subsequent decline. While trade and industry can generate spurts of growth they do not, on their own, guarantee long-term economic viability. For cities to survive and grow over the long term they need to be large and have a diversified economy. Size does matter! The urban economist Wilbur Thompson (1965) identifies what he calls a "ratchet effect": above a critical size, cities will not decline. Thompson cites four reasons: larger cities have a more diversified base so that growth will be maintained even if a particular sector declines; they wield more political muscle and thus put more claims on public expenditures to stave off shrinkage; the big city is an important market in its own right; and big cities are more likely to be sites of innovations which provide the basis for subsequent growth. He proposed that the critical threshold figure was a population of one-quarter of a million, but it is better to consider the threshold as a relative figure varying over time and space. Whatever the exact amount, it is clear that the bigger the city, the more the ratchet effect comes into operation. But as we shall see, even large cities (of almost 1 million people), such as Detroit and Baltimore, have witnessed marked decline in the face of manufacturing loss. The most extreme examples, ghost towns, are testaments to the fragility of economic growth based on single industries in small towns.

In terms of a narrow reliance on trade, regional systems may shrink to reliance on a few centers. Allen Pred (1966) examined the mercantile cities along the eastern coast of the US in the period 1800 to 1840. In 1800 there was a large number of small cities, including Charlestown, New Bern, Norfolk, Baltimore, Trenton, Milford, New London, Gloucester and many more. As trade grew there was an expansion of the wholesale trading system involving the construction of warehouses that involved the recruitment of labor, which increased local purchasing power, resulting in higher demand for goods and services. The process of cumulative causation in which growth promotes more growth was felt in a select group of more advantageously positioned cities. By 1840, Charlestown, Baltimore, Philadelphia, New York and Boston dominated the trading system and were centers of urban growth, while New Bern and Milford shrank in importance.

In terms of industrial growth, we can picture a three-stage model. In the first stage, firms locate on greenfield sites, labor is attracted and population increases. In the second, more mature stage local firms make profits, wages rise and there are buoyant housing and labor markets. However, as technological developments render certain processes and plants less efficient, capital disinvests to seek more profitable locations; a period of decline is inaugurated that involves the out-migration of skilled workers as well as reduced spending power and employment opportunities. The process of cumulative causation is now reversed as a spiral of decline feeds into more decline. In urban economies where job losses are not offset by growth in other sectors, the whole city declines.

A buoyant urban center is reliant on a large and diversified economy. Remember the example of Schenectady, where the loss in manufacturing employment was not offset by job growth in other sectors to halt a downward spiral. There are also many counter-examples. For decades, Hong Kong was a center of manufacturing, one of the original Asian "tigers" of rapid economic growth in the 1960s and 1970s. However, there was a steady decline in the importance of manufacturing. In 1980, almost one out of every two workers was involved in manufacturing: by 1997 this had fallen to fewer than one in ten. But the decline was offset, in total employment terms, by an increase in producer services, especially manufacturing producer services involved in servicing import–export, financing and real estate (Tao and Wong, 2002). Hong Kong became an important center for servicing the manufacturing sector that had shifted to mainland China. This sort of shift from manufacturing to service is the hallmark of successful urban economies.

The growth of the service sector

In mature economies there has been shift from manufacturing to services. In the US, for example, service employment now accounts for one in every three US workers and almost 30 percent of GNP. Services are defined as selling assistance and expertise rather than a tangible product. In some cases the distinction is clear: when you buy a car, you are buying a product; when you hire someone to clean your house, you are buying a service. But at other times the distinction is fuzzy: the classic example is a restaurant, where you buy both food *and* service. "Services" is best considered as a loose and hazy term that at the edges slips over into goods. It also covers a wide range of activities, from healthcare and financial consultancy to computer information companies. The sector includes a range of wildly differing job experiences. At one end are the highly-paid Wall Street brokers working in international currency dealing whose lucrative salaries and bonuses fuel local housing markets. At the other end are contract cleaners who tidy up after these executives (Aguiar and Herod, 2007).

One particularly significant part of the services sector comprises the knowledge-based industries, so-called "producer services," such as advertising, banking services, financial services, business consultancies and information technology. Together these businesses constitute the dynamic edge of mature capitalist economy. Since 1980, in the developed world, a city's success rests less on manufacturing employment and more on the extent to which it can generate, retain and attract such services. In the nineteenth and early twentieth centuries the factory contained both the assembly line and the offices that administered the whole process of buying raw material, hiring workers and selling finished products. The business service sector has become more prominent as these services sectors have been hived off into separate divisions and into separate companies. Take, for example, a simple model of a contemporary multinational manufacturing business. At the base is the routine assembly plant which needs all those things noted in the standard location models of industry: cheap labor, low taxes, etc. Because of the ease and low cost of transporting goods, locational constraints are less powerful, so these plants can be located in a range of countries around the world. Economic globalization has therefore created a flatter world. Another level of the company that has become separated from the production plants is the research and development sector, which tests new products. This sector needs to be close to pools of highly skilled labor, knowledge pools and the amenity-rich locations that attract such workers. Then there is the company headquarters, which needs a metropolitan location in order to maximize face-to-face business contacts and be close to business services that provide advertising, financing, legal services and other important services. Consider the case of Boeing, a company with almost 156,000 workers around the world, almost 66,000 of them in the state of

Case study 4.1

The 20 largest cities in the US, 1850–2000

Table 4.1 reveals some of the trends we have discussed in the previous three chapters. First, large and economically successful cities such as New York maintain their primacy. From 1850 to 2000 New York City saw an increase and then a decrease in the amount of manufacturing employment. But the decline was offset by an increase in service employment, especially in the advanced producer services that require a central city location. A similar trend can be noted, albeit time delayed, in the growth and decline of manufacturing for Chicago and later Los Angeles.

A second trend is the relative, and indeed absolute, decline of the older industrial cities of the Northeast. Baltimore, for example, was the second-largest city in 1850, maintaining its position in the top ten until 1970. Thereafter, massive deindustrialization with limited offsetting service employment increase led to a decline. By 2000 Baltimore stood at number 18. Cincinnati was the eighth-largest city in 1880 but had fallen out of the top 20 completely by 2000. Similar trends can be noted for other industrial cities. In 1970, Detroit, Milwaukee and Cleveland were fifth, twelfth and tenth, respectively, but by 2000 they were tenth, nineteenth and no longer in the top 20.

A third trend is the growth of urban centers in the Sunbelt region caused in part by industrial relocation, and more recently by new job formation. In 1880 Phoenix, San Diego, San Antonio, Dallas and San Jose did not figure; by 1970 they had squeezed into the top twenty; and by 2000 they were firmly established as large US cities.

To some extent suburbanization, the shift from the central cities to suburban areas that are separate municipalities, explains some of the decline of such cities as Atlanta. Metropolitan fragmentation is an urban reality in the US. However, the changes have been shaped by, and embody, broad-scale economic changes and in particular the rise and fall of industrial cities in the Northeast, the growth of new cities in the Sunbelt and the enduring dominance of the very large successful urban economies of New York, Chicago and, more recently, Los Angles. The changing fortunes of cities in the Northeast and the Sunbelt are clearly shown in Figure 4.1. In 1900, all large American cities, except for San Francisco and New Orleans, were located in the Northeast. Just one hundred years later, only seven Northeast cities remained within the top 20. In the meantime, California and Texas each boasted four cities in the list. Population rank is a good barometer of changing urban economic fortunes.

Table 4.1 The 20 largest cities in the US, 1850–2000

Rank	1850	1880	1910	1940	1970	2000
1	New York	New York	New York	New York	New York	New York
2	Baltimore	Philadelphia	Chicago	Chicago	Chicago	Los Angeles
3	Boston	Brooklyn	Philadelphia	Philadelphia	Los Angeles	Chicago
4	Philadelphia	Chicago	St. Louis	Detroit	Philadelphia	Houston
5	New Orleans	Boston	Boston	Los Angeles	Detroit	Philadelphia
6	Cincinnati	St. Louis	Cleveland	Cleveland	Houston	Phoenix
7	Brooklyn	Baltimore	Baltimore	Baltimore	Baltimore	San Diego
8	St. Louis	Cincinnati	Pittsburgh	St. Louis	Dallas	San Antonio
9	Spring Garden	San Francisco	Detroit	Boston	Washington, DC	Dallas
10	Albany	New Orleans	Buffalo	Pittsburgh	Cleveland	Detroit
11	Northern Liberties	Cleveland	San Francisco	Washington, DC	Indianapolis	San Jose
12	Kensington	Pittsburgh	Milwaukee	San Francisco	Milwaukee	Indianapolis
13	Pittsburgh	Buffalo	Cincinnati	Milwaukee	San Francisco	Jacksonville
14	Louisville	Washington, DC	Newark	Buffalo	San Diego	San Francisco
15	Charleston	Newark	New Orleans	New Orleans	San Antonio	Columbus
16	Buffalo	Louisville	Washington, DC	Minneapolis	Boston	Austin
17	Providence	Jersey City	Los Angeles	Cincinnati	Memphis	Memphis
18	Washington, DC	Detroit	Minneapolis	Newark	St. Louis	Baltimore
19	Newark	Milwaukee	Jersey City	Kansas City	New Orleans	Milwaukee
20	Southwark	Providence	Kansas City	Indianapolis	Phoenix	Fort Worth

Source: US Census Bureau (1998 and 2006)

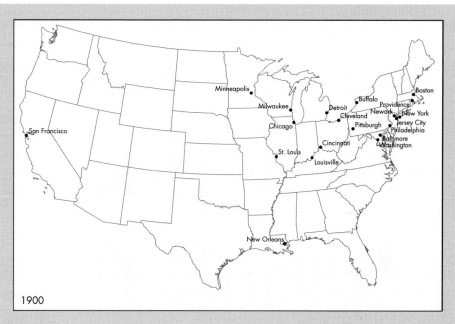

1900

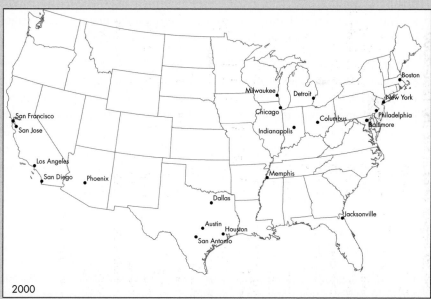

2000

Figure 4.1 Geographical shift of the 20 largest cities in the US, 1900 and 2000

Washington. In 2001 the company moved its headquarters from Seattle to Chicago and separated its corporate headquarters from its production base. According to the company, the move was made in order to secure ready access to global markets and easier access to financial markets. It was also lubricated by subsidies of over $60 million from the state of Illinois and the city of Chicago, which was in competition with Denver and Dallas–Fort Worth.

Financial services remain concentrated especially in the global cities, such as New York, where the economics of agglomeration and the ability to provide face-to-face contact are of great importance. Gehrig (1998) outlines the various forces shaping financial centers. The centripetal forces include economies of scale, information spillover, liquid markets and labor markets. Centrifugal forces include market access costs and localized information. The former forces dominate, and the current structure is a network of cities that services the contemporary transnational and securitized form of financial systems in a strongly hierarchical form. At the apex are the supranational centers of London and New York, followed by a second tier of international centers that include Tokyo and Zurich, then host centers which attract foreign financial institutions, such as Sydney, Toronto and Vienna. The determinants of financial center competitiveness include agglomeration of demand and supply, a culture of information expertise and contacts, avoidance of overregulation, well-run and disciplined markets, and ongoing innovation that promotes rapid dissemination and quick response.

The globalizing economy creates lots of information, narrative uncertainty and economic risk that all have to be produced, managed, narrated, explained and acted upon. Global cities are centers of global epistemic communities of surveillance, knowledge production and storytelling (Storper, 1997). Trust, contact networks and social relations play pivotal roles in the smooth functioning of global business. Spatial propinquity allows these relations to be easily maintained, lubricated and sustained in an efficient means of communication that helps solve incentive problems, facilitates socialization and learning, and provides psychological motivation. Storper and Venables (2004) write of the positive benefits of face-to-face contact in the urban economy. Global cities are the sites of dense networks of interpersonal contact and centers of the important business social capital trust vital to the successful operation of international finance.

There are a number of empirical studies of individual sectors of advanced producer services that highlight their global city bias. Beaverstock et al. (2000) assess the geographical spread of the foreign offices of US law firms. Just 15 cities house 73 percent of the total, with London alone accounting for 17 percent of all offices of US law firms abroad. And 59 percent are found in just eight cities: London, Hong Kong, Paris, Tokyo, Brussels, Moscow, Singapore, and Frankfurt.

The US law firms that had a global presence were concentrated in New York, Chicago, Washington, DC, Philadelphia, Boston and Los Angeles. The same authors had earlier looked at the location of London's law firms (Beaverstock *et al.*, 1999). All but 1 of the top 30 London law firms had a foreign presence, that in total comprised 221 foreign branches in 60 cities. Brussels was the most dominant city, with 25 offices, followed by Hong Kong (18), Paris (13), Singapore (12), and New York (10). (The dominance of Brussels reflected the need for proximity to the EU Commission and its bureaucracy.)

Global service corporations depend upon specialized knowledge. Sassen (1994) suggests that global cities are knowledge-rich environments, and that face-to-face contacts between experts are facilitated by the clustering of knowledge-rich individuals in cities like New York, London and Paris. In this way, global cities have become "privileged sites" in the contemporary world economy, housing the "knowledge elite" who act as crucial mediators and translators of the flows of knowledge, capital, people and goods that circulate in the world. A global city attends to the heterogeneous global space of flows, lending otherwise incommensurable materials intelligibility and translatability.

Researchers in the Globalization and World Cities (GaWC) group examined the distribution of advanced producer services across a range of cities. They generated a data matrix of 316 cities and 100 firms in accountancy, advertising, banking/insurance, law and management consultancy; identified firms with at least 15 separate offices; and analysed connectivity between the 316 cities. Those that had at least one-fifth of the connectivity of the most connected city (London) were identified as world cities. A total of 123 world cities were identified in this way (see Taylor, 2004). The global urban system they identify is a network of – and for the functioning of – advanced finance capitalism. If we had chosen another set of flows – for example, flows of migrants – a very different network would have been identified (Benton-Short *et al.*, 2005).

The growth of the service sector also produces a more polarized job market with, on the one hand, high-paying jobs with full benefits and, on the other, minimum-wage jobs with few benefits. At the core of the new service economy are the highly paid knowledge-based professionals, symbolic analysts such as business consultants and investment bankers. Paid generously, they have employment security, good working conditions and generous benefits. They often work long hours with brutal deadlines, but they are firmly located in the middle to upper-middle class, and they can assure their children's similar economic success by buying good private education or living in residential areas that provide high-quality public education. At the very upper levels of this sector is a world of affluence that is often represented but rarely experienced. It is the world of the corporate jet and the million-dollar-plus stock options that provide a deep financial cushioning from the vagaries of life. Outside of this core group are two others. One, on the semi-periphery, consists of full-time workers with less income, status and prestige. Their jobs are less secure, and often they hang on to middle-class status only by having more than one person in the household working. On the true periphery comprises people on short-term or part-time contracts. They may work from home on computers and relish the flexibility of such work or may be picked up on a daily basis on the street corner by landscape gardeners to manicure the lawns of the wealthy.

The shift from a manufacturing-based economy to a service-based economy has profound effects on income distribution. A strong manufacturing base allowed low-skilled workers to obtain relatively good wages. The service economy, in contrast, provides high-paying jobs for those with marketable skills, but more limited opportunities for low-skilled workers. For those lacking high educational attainment, opportunities exist only in the low-waged service sector.

The shift from manufacturing to a more service-orientated economy also goes hand-in-hand with an increase in female employment. In the US in 1950 female participation rate in the formal economy was only 30 percent; by 2001 it had increased substantially. Out of a total employed labor force of 135 million, 63 million were now women. The workforce had become feminized.

The role of the public sector

Public employment is a form of service employment that is an important sector of urban economies. At the federal, state and local levels, there has been steady increase in the absolute and relative share of public employment. Since the public sector tends to be more unionized than the private sector, public sector employment acts as a counterweight to the decline of union membership in the private sector. Public sector employees in the US, for example, are four times more likely to be in a union than private sector employees. The heavy unionization of the public sector, along with relatively good working conditions and employment security, reinforces the outsourcing of work as cash-strapped governments seek cheaper, more pliable workers.

Public employment is a significant element in city finances and urban politics. The increasing size of city payrolls is in part a function of rising demand for municipal services. However, it is not a simple case of municipal employment growing in line with rising needs. There is also pressure from public sector unions to win jobs and better conditions, and mounting political claims from urban constituencies eager to benefit from municipal largesse. Frances Fox Piven and

Richard Cloward (1997) make a convincing claim that the urban fiscal crisis of the 1970s and 1980s in the US was caused in part by cities being unable to raise revenues yet equally unable to resist the claims that underlie rising expenditures. The crisis in part was caused by the rising demands of public sector unions and political constituencies wanting to gain access to secure public employment. Municipal employment is part of the political compromise worked out by business elites and community leaders to maintain political peace and lubricate the workings of urban regimes. It is also a part of the conflict and compromise in the urban political arena. We can identify a number of diverse interests, including those of taxpayers, users of services and municipal workers. These groups are not mutually exclusive: some people fit in all three. Parents of schoolchildren complain of poor-quality education; but the same people, as taxpayers, baulk at rising tax bills; and they also may be teachers who want to keep their jobs and see more teachers employed to lighten their load. There are obvious sources of conflict between the groups. Users of services want efficient, good-quality services. Taxpayers want their taxes kept low. Municipal workers want protected, high-paying jobs. In some cases all of these interests can be satisfied, but in many other cases tensions bubble up into direct conflict. Municipal employment is not just another job category; it is an important element in the ongoing struggle in the urban political arena.

Global shift of services

From the early 1970s global shift mostly affected the manufacturing sectors. In more recent years there has been a significant increase in the global shift in service employment. A new round of economic globalization, made possible by changes in technology, is sending a range of service employment overseas from the developed world to the developing world. Back offices in Bangalore, India, now process home loans for US mortgage companies while many insurance claims made in the US are routinely processed in offices situated in New Delhi. The economics are simple. Software designers in the US cost $7,000 a month while experienced designers in India cost only $1,000 a month. US companies now routinely outsource work previously done only at home. In the 1970s and 1980s engineers would come to the US and Europe; now the jobs come to them.

Selected cities in developing countries are developing as centers of service employment. In part they develop as hubs of corporate national headquartering, which generate service employment. In China, for example, Beijing and Shanghai are the favored sites for headquarters of foreign companies. Zhao's analysis of data from 2000 shows that 46 percent of foreign service companies in China establish their base in Beijing and 27 percent in Shanghai. In the financial services sector the figures are 44 percent and 30 percent; while for advertising and business consultancy they are 53 percent and 27 percent (Zhao, 2003). In China corporate companies are attracted to Beijing because of the closer proximity to key decision-makers, an important criterion to work effectively in a centrally planned economy. However, even in the secondary city of Shanghai there has been a shift from manufacturing to service employment. In 1995 there were almost 400,000 people employed in manufacturing and 300,000 in service. By 2002 manufacturing employment had shrunk to 288,000 while service had increased to 313,000. While only one in four jobs in China is classified as service, almost one in two in Shanghai is so classified. By 2003 the financial services sector alone accounted for 10 percent of the city's GDP.

We can summarize this chapter very simply. Cities can develop on the basis of trade and, since the late-eighteenth century, on manufacturing. However, changes in consumption, technological developments and the global shift in manufacturing can all undermine urban economies based on these activities. Urban economic history is studded with the rise and fall of merchant cities and industrial cities. In order to have long-term and sustainable growth urban economies need a broad and diverse economic base. Service employment provides a broader base with advanced producer service, in particular, as the leading sector of rapid urban growth. Over the past thirty years we have seen the decline of manufacturing employment and a rise in service employment. This has been a pervasive trend throughout the cities of the developed world and is now increasingly evident in selected cities of the developing world, too. Around the world the more dynamic urban economies are shifting from manufacturing to services.

Urban regimes

Urban economies and urban economic change are closely tied to local politics. The relationship has been considered in four interrelated ways. First, Logan and Molotch (1987) identify what they call the "urban growth machine" in US cities consisting of realtors, local banks, influential politicians, corporate chairs and chambers of commerce which constitute a lobby that persuades city politicians to concentrate on stimulating investment and economic growth. The lobby therefore promotes a pro-business agenda for their particular city. The increasing com-petition between cities can strengthen this growth machine. Jonas and Wilson (1999) provide a more recent assessment of the urban growth machine.

Second, Clarence Stone (1989: 3), drawing on his analysis of the governing of Atlanta from 1946 to 1988, identifies "urban regimes" that he defines as the "informal arrangements that surround and complement the formal workings of governmental authority." Urban regimes consist of informal governing coalitions that make decisions and get things done in a city. Where Logan and Molotch concentrate on the economic agenda of the business elite, Stone highlights their interactions with political power and the resultant compromises. Political questions of maintaining and extending political support and leadership dominate City Hall, while economic issues of profit and loss concern the business elites. The combination of political and economic logic, with all the ensuing tensions, conflicts and ambiguities, constitutes the local urban regime. Urban regime analysis has stimulated a great deal of work (Mossberger and Stoker, 2001; Stone, 2005).

Urban regimes vary over time and space. Judd and Kantor (1992) identify four cycles of regime politics in the US. In the *entrepreneurial cities* up to the 1870s merchant elites controlled the city. Then, with industrialization and large-scale immigration, business interests had to work with political representatives of the newly organized immigrants. The result was the *city of machine politics* that saw its high point from the 1870s to the 1930s. From the 1930s to the 1970s a *New Deal coalition* prevailed in which federal policies stimulated urban economies and maintained the Democratic power base. In the contemporary cycle the regime promotes *economic growth and a political inclusiveness*.

Stoker and Mossberger (1994) identify three regime types: organic, instrumental and symbolic. *Organic regimes* occur in small towns and suburban districts with a homogeneous population and a strong sense of place; their chief aim is to maintain the status quo. *Instrumental regimes* focus upon specific targets identified in the political partnership between urban governments and business interests. *Symbolic regimes* occur in cities undergoing rapid changes, including large-scale revitalization, major political change, and image campaigns that try to shift the wider public perception of the city. These are ideal types, with any one city's regime capable of exhibiting characteristics of each type.

Central city regimes, such as in Atlanta, consist of alliances of business interests and the power brokers of the formal political machines. Urban growth issues merge with issues of social inclusiveness, and the need to maintain and extend political support across a coalition of varied political interests tempers the pro-growth lobby. In the suburban regimes a shared agenda of keeping taxes low and protecting property values unites the population. Here the local regimes exist to maintain the status quo and often to keep out lower-income and racially different groups.

Third, emphasis has been placed on the reasons for and consequences of the regime change from the Keynesian–New Deal city to the entrepreneurial city (Harvey, 1989a; Hall and Hubbard, 1998). The Keynesian city is named after the British economist John Maynard Keynes (1883–1946), who argued that government had a major role to play in stimulating effective demand in the economy. The period from 1933 through to the 1980s marks the high point of the Keynesian–New Deal city in much of the developed world, when there was a consensus between capital and labor on the role of government. Government spending stimulated demand so that unemployment would be limited and controlled. Government-funded programs ensured that the majority of the population had access to relatively affordable health, housing, education and social welfare, which all softened the social consequences of business downturns. In the US business interests held a stronger hand in comparison to Northwest Europe, where organized labor was more powerful and social welfare programs were not so curtailed by significant resistance to taxes and to the role of government in general. From the 1980s the Keynesian–New Deal city began to disappear because of: the persistence of "stagflation" that seemed to disrupt the balancing act of government spending that could minimize unemployment while avoiding inflation; growing resistance to government taxation as programs were funded by deepening and widening the income tax and local property tax base; increasing economic competition and the declining power of organized labor. Beginning in the 1980s a new meta-narrative took over, often referred to as the neo-liberal agenda, that limited government spending, especially on welfare programs, reduced social subsidies, freed up markets, globalized economies, imposed limits on tax increases, all resulting in a massive redirection of government spending and a dramatic reorientation in the nature of national and city politics.

The neo-liberal agenda creates the entrepreneurial city that seeks to "facilitate privatization and the dismantling of collective services" in order to take advantage of the opportunities of connecting with the global economy (Lauria, 1997: 7). In the entrepreneurial city urban regimes are more concerned with direct income generation and a variety of public–private partnerships. Cities have become more concerned with the politics of maximizing growth and income than with their redistribution (Hall and Hubbard, 1998).

Fourth, particular attention is being paid to how urban regimes are adapting and responding to globalization and increased global competition. Short (2004a) suggests that the focus should be shifted to the political agendas of globalizing cities and points attention to the representation of industrial cities, the increased use of global spectacles and signature architects and the creation of both a cosmopolitan urban semiotics and an explicitly pro-business climate. The city is now a place of global imaginings as cities compete for global recognition and business, and existing global cities struggle to maintain their positions. A more complex discernment of the role of urban regimes in the political creation of global and globalizing cities will allow us to understand how local and global economic forces interact.

Further reading

Bell, Daniel, 1973, *The Coming of Post-Industrial Society*, New York: Basic Books. A path-breaking work that first coined the term "post-industrial." A speculative form of social forecasting that accurately predicted economies more based on the economics of information and the production of knowledge. Dated but a classic work.

Daniels, Peter, Andrew Leyshon, Mike Bradshaw and Jonathan Beaverstock, eds, 2007, *Geographies of the New Economy: Critical Reflections*, London: Routledge. A very good collection of papers that draw on the experience of cities around the world to consider global shift, post-industrial growth, the development of service economies and the role of advanced producer services.

Ehrenreich, Barbara, 2001, *Nickel and Dimed: On (Not) Getting by in America*, New York: Metropolitan. This is a well-written, first-person, participant account that tells the story of those at the wrong end of the service economy. The author worked as a waitress, hotel maid, nursing-home aide and sales clerk. Whatever the job, it was a depressingly similar story of little pay for many hours of hard work, with few benefits and punishing schedules. An inside look at the lower level of the US service economy.

Reich, Robert B., 1991, *The Work of Nations*, New York: Knopf. An engaging and accessible argument that highlights the shift from high-volume to high-value production, the footloose nature of corporations and the increasingly important role of symbolic analysts. As much political tract as economic analysis and all the better for that.

Part Two

The global economy and world cities in developed countries

Urban economic change has been closely linked to a series of historical processes, including the emergence of capitalism, industrialization, modernization and deindustrialization. The process of globalization has recently joined the list. Part Two examines the relationship between globalization and urban change in large cities in the developed world.

The effects of globalization on people and on the world, in all forms and at all levels, have been an underlying theme in globalization studies. However, some urban scholars have successfully conceptualized recent economic, cultural, political and spatial changes in large cities not just as outcomes but as enabling factors of globalization (Sassen, 1991). This notion appears in numerous case studies of leading world cities, such as London, New York and Tokyo, where the so-called "movers and shapers" of the current world economy are clustered. However, the impact of globalization in other, more ordinary cities, or non-world cities, has received much less scholarly attention. We attempt to examine the impact of globalization on both world cities and non-world cities in the three chapters of Part Two.

One recurring debate about the urban economy concerns what city governments can do to ensure long-term growth. This issue of development policy has come to the fore again recently, particularly with regards to the devastating effects of globalization in traditional industrial cities. Such effects include deindustrialization and full-blown urban decline. No one can say for sure which policies will most effectively improve urban economic conditions, yet that has not stopped city officials from claiming that "their" development strategies will deliver economic success and even world-city status. In this part, we pay particular attention to the political rhetoric of urban economic development, as well as to the reality.

In Chapter 5, we analyze the relationship between globalization and world cities, which are at the apex of the global urban hierarchy. Chapter 6 examines the impact of globalization on cities below the category of world city, with particular focus on the politics of world-city status and urban competitiveness. In Chapter 7, we look into new ideas and policies for urban economic development, including the cultural economy and the creative, knowledge-based economy – both often deemed critical to urban success in the era of globalization.

5 Globalization and world cities

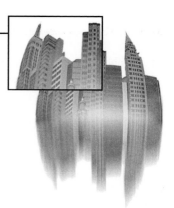

Learning objectives

- To think about the relationship between globalization and urbanization
- To inquire what constitutes a world city in the contemporary world
- To have a critical understanding of world-cities research

The interface between globalization and urbanization pervades contemporary urban studies. External factors have long affected cities, yet the geographical scale of causes, processes and outcomes of urban changes has grown increasingly transnational, especially in recent decades. The fact that so many large cities, despite their distinctive histories and socio-political systems, are experiencing similar economic, cultural and spatial changes lies at the heart of so-called "globalization–urbanization nexus" literature. In many of these studies, globalization is interpreted not only as a major source of urban change but as a process that is itself facilitated by these changes.

The impact of globalization has been observed in various urban sites around the world, yet the limited number of so-called world cities or global cities – namely London, New York and Tokyo, and, to a lesser degree, Los Angeles and Paris – have received the majority of academic scrutiny. World cities are those that function as primary hubs for global networks of business firms, financial institutions, (non-)governmental organizations and migrants – the "central places where the work of globalization gets done," according to Sassen (2002: 8). Some argue that globalization simultaneously causes and reflects changes in the economies, cultures, politics and geographies of world cities. This idea reinforces the role of world cities – not nation states – as basic units of analysis for globalization. Taylor (2004) even promotes a shift from the state-centric to the city-centric view of the world for a deeper understanding of globalization and the unfolding world system.

In this chapter, we review the existing studies of world cities to assess what has been, or has not been, agreed on the list of world cities and their particular role in globalization. The first section reviews world-cities research in urban studies, which dates back to the early twentieth century. In the second section, we look at the leading world cities that seem to command the current world economy most effectively. The third and final section critically examines the data sources and analytic methods used to draw the global hierarchy of world cities.

World cities in urban studies

The term "world cities" was coined by Patrick Geddes in *Cities in Evolution* (1915) to illustrate urban growth and conurbations in city-regions outside of Great Britain. In this book, Paris, Berlin and New York were noted for their growing suburbs, while other fast-growing cities, such as Düsseldorf and Pittsburgh, also drew Geddes's interest.

The term was reintroduced and given a new meaning in Peter Hall's *The World Cities* (1966). Hall identified seven world cities – London, Paris, Randstad (The Netherlands), Rhine–Ruhr, Moscow, New York and Tokyo – which owned and conducted disproportionate shares of the world's most important businesses. He pointed to political power and trade, including transportation, banking and finance as factors distinguishing the world cities from other great centers of population and wealth. In the 1960s, as the formation and evolution of national urban systems dominated urban researches (see Berry, 1964; Borchert, 1967; Pred, 1966), Hall's work initially helped to identify the most politically and economically powerful cities on an international scale. However, he focused more on their individual growth and problems than on the connections or competition among them.

Discussion of world cities exploded with John Friedmann's now-classic essay "The world city hypothesis" (1986). Friedmann's argument built upon ideas formulated by Cohen (1981) and Friedmann and Wolff (1982) that situated major cities within the international division of labor. (Barlow and Slack (1985)'s work on "international cities" was not cited in Friedmann's works, although they pointed out a neglect of the international context in urban studies and argued for more researches on the role of "international urban activities" in economic changes in major cities.) Friedmann called for a new research framework in which urbanization processes are linked to global economic forces. Within that framework, he defined world cities as: basing points for global capital in the spatial organization and articulation of production and markets; sites with expanding sectors attached to corporate headquarters, international finance, global transport and communications and high-level business services; major sites for the concentration and accumulation of international capital; and points of destination for large numbers of domestic and/or international migrants. Based on these criteria, he identified thirty world cities and put them in four hierarchical categories (Table 5.1). His original list of first-tier world cities includes London and Paris in Western Europe; New York, Chicago and Los Angeles in North America; and Tokyo in Asia.

In a subsequent article, "Where we stand: a decade of world city research" (1995), Friedmann proposed a slightly changed global urban hierarchy, with only London, New York and Tokyo at the top. Below these top global financial articulations, he identified multinational articulations (Miami, Los Angeles, Frankfurt, Amsterdam and Singapore), important national articulations (Paris, Zurich, Madrid, Mexico City, São Paulo, Seoul and Sydney) and subnational or regional articulations (Osaka–Kobe, San Francisco, Seattle, Houston, Chicago, Boston, Vancouver, Toronto, Montreal, Hong Kong, Milan, Lyon, Barcelona, Munich and Rhine–Ruhr) (Plate 5.1). Although some have suggested modifications to his list of world cities (for example, Lo and Yeung, 1996; Short *et al.*, 1996; Taylor, 2000),

Table 5.1 Friedmann's world city hierarchy

	Europe	Americas	Asia	Rest of the world
Core:				
primary city	London	New York	Tokyo	
	Paris	Chicago		
	Rotterdam	Los Angeles		
	Frankfurt			
	Zurich			
secondary city	Brussels	Toronto		Sydney
	Milan	Miami		
	Vienna	Houston		
	Madrid	San Francisco		
Semi-periphery:				
primary city		São Paulo	Singapore	
secondary city		Buenos Aires	Hong Kong	Johannesburg
		Rio de Janeiro	Taipei	
		Caracas	Manila	
		Mexico City	Bangkok	
			Seoul	

Source: Fridemann (1986)

Plate 5.1   Hong Kong's claim: Asia's world city

Friedmann's world-city hypothesis has held up over the past two decades as a leading point of reference for both empirical and theoretical studies of cities in a globalizing economy (Figure 5.1).

Friedmann's insistence that worldwide processes affect urban change inspired many scholars of urban studies to take a global perspective (Brenner and Keil, 2006). Even though third world urbanization was repeatedly contextualized in international political and economic processes in the 1970s and 1980s, prior to Friedmann's work the global framework was rarely used to investigate the form and function of major cities in the developed world, particularly North America (Davis, 2005). Most earlier studies on intercity relations, such as analyses of central place, urban systems, spatial diffusion and transportation network, set urban centers on the national or even sub-national scale.

Friedmann's notion of a world-city system has contributed immensely to a better understanding of large metropolises and their interconnections by setting its object of analysis at a global scale. Ever more cities today are tightly embedded within various global networks, such as multinational firms' production and marketing

AMERICAS

EUROPE–AFRICA–
MIDDLE EAST

ASIA–OCEANIA

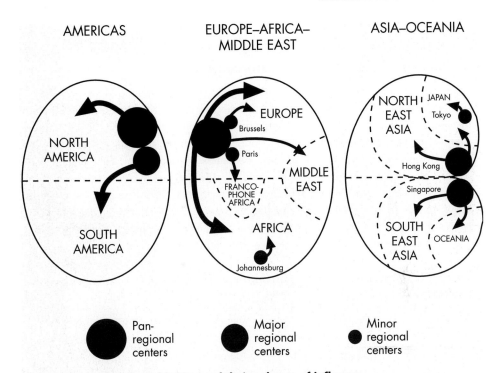

Figure 5.1 **Regional world cities and their spheres of influence**

Source: **Taylor (2000: 25)**

networks, international labor migration flows, and the worldwide web of information resources. Accordingly, the geographical scale of the causes, processes
and outcomes of urban changes is now much more transnational than ever before.
Certain cities, such as Dubai and Miami, stand out for their role in facilitating
transnational flows of people, capital and cultures, rather than for their domestic
presence.

While the notion of world cities as a research framework has been well accepted
among urban scholars, the research methods that Friedmann and his followers
(collectively called the "World City School") used to rank cities have been
questioned extensively. We examine three of the major criticisms below.

First, while the interplay of globalization and urban change affects many aspects
of urban society, most studies of world cities focus narrowly on economic change.
Many have argued that, along with a cluster of leading multinational corporations,
financial firms and business services, cosmopolitan cultures and immigrant
communities represent a key feature of world cities (Abrahamson, 2004; Benton-
Short *et al.*, 2005; Hannerz, 1996).

Case study 5.1
American cities, cosmopolitan cities

Immigrant communities, along with ethnic neighborhoods, have long been a familiar scene in many large American cities. According to the US Census Bureau's 2005 American Community Survey, the foreign-born population accounts for 35.7 million people, or 12.5 percent of the total US population. Among the foreign born, 53.3 percent were born in Latin America (30.7 percent in Mexico alone), 26.7 percent in Asia, 13.6 percent in Europe, and the remaining 6.4 percent in other regions of the world. A disproportionate share of these new immigrants live in metropolitan areas, 44.4 percent in central cities and another 50.3 percent outside central cities but still within a metropolitan area. Compared to the native-born, immigrants demonstrate a clear urban orientation, often preferring certain cities.

Table 5.2 **American cities with highest percentages of foreign-born residents**

Rank	City	Percentage of foreign-born, 2004
1	Miami, FL	58.7
2	Santa Ana, CA	55.4
3	Los Angeles, CA	40.4
4	San Jose, CA	37.9
5	San Francisco, CA	37.3
6	New York, NY	35.9
7	Anaheim, CA	35.4
8	Long Beach, CA	31.2
9	Stockton, CA	29.9
10	Oakland, CA	27.2
11	San Diego, CA	27.1
12	Houston, TX	27.0
13	Boston, MA	26.8
14	El Paso, TX	26.7
15	Dallas, TX	26.6
16	Honolulu CDP, HI	26.4
17	Riverside, CA	24.0
18	Newark, NJ	23.8
19	Sacramento, CA	23.6
20	Phoenix, AZ	22.8
21	Chicago, IL	21.3
22	Las Vegas, NV	20.8
23	Seattle, WA	20.8

Source: US Census Bureau (2006)

Table 5.2 lists twenty-three American cities in which at least one in five residents was not a US citizen at birth. In Miami, foreign-born residents account for almost 60 percent of its population, with many of the new arrivals from Cuba. Cities in California, such as Los Angeles, receive a large number of Mexican immigrants, while New York has been known for its diverse population, including migrants from the Dominican Republic, China, Jamaica and Guyana.

The recent surge in Latino migration has targeted cities in the Southwestern states, such as Long Beach in California. This reflects a very new trend in the relationship between immigration and urban growth in the US. Table 5.3 illustrates the dynamics of US urban immigrant history. New York and Los Angeles have received a steady stream of foreign-born migrants over the years, though a series of Immigration Acts have varied its intensity over time. Industrial cities in the Northeast, like the two Ohio cities Cincinnati and Cleveland, once recruited large numbers of foreign-born workers, but as their industrial bases declined, immigrant arrivals decreased as well. It is not clear whether economic decline caused immigration decline or vice versa. Cincinnati was once a major destination of German and, in the nineteenth century, Jewish immigrants. Now its foreign-born population accounts for less than 5 percent of the total. Cleveland still

Table 5.3 Changes in the percentage of foreign-born residents in selected US cities

	New York	Los Angeles	Long Beach	Cleveland	Cincinnati
1850	45.7	NA	NA	NA	47.2
1860	47.2	NA	NA	NA	45.7
1870	44.5	35.0	NA	41.8	36.8
1880	39.7	28.7	NA	37.1	28.1
1890	42.2	25.3	NA	37.2	24.1
1900	37.0	19.5	NA	32.6	17.8
1910	40.8	20.7	NA	35.0	15.6
1920	36.1	21.2	12.8	30.1	10.7
1930	34.0	20.0	10.1	25.6	7.8
1940	28.7	15.1	7.7	20.5	5.7
1950	23.6	13.4	6.5	14.6	4.1
1960	20.0	12.6	6.2	11.0	3.3
1970	18.2	14.6	6.8	7.5	2.7
1980	23.6	27.1	14.2	5.8	2.8
1990	28.4	38.4	24.3	4.1	2.8
2004	35.9	40.4	31.2	4.1	4.3

Source: US Census Bureau (1999 and 2006)

boasts distinctive Eastern European ethnic neighborhoods, but its immigrant population has shrunk dramatically in the past few decades as well.

In recent years, the high number of illegal immigrants, estimated at over 10 million, has created a heated political and public debate regarding their legal status and contribution to the American economy. Largely absent from the debate has been analysis of their contribution to cosmopolitanism and multiculturalism in large American cities. Along with world-class corporate executives, bankers, stock dealers, lawyers, consultants and artists, the large immigrant population and the cultural diversity it offers have made New York an outright world city.

Second, some worry that world-cities research perpetuates a geographic bias toward exceptionally large metropolises in the Western world, leading to a lack of understanding of other cities (McCann, 2004; Robinson, 2002; Gugler, 2004). Indeed, both smaller cities in developed countries and almost all cities in developing countries have been overlooked, even as many of these cities integrate increasingly into a global economy.

Finally, others sharply criticize world-cities research's methodological bias toward the hierarchical categorization of cities. While many scholars identify attributes separating the leading world cities from the second- or lower-tier cities, few address the nature of linkages among world cities (Smith and Timberlake, 1995; Taylor, 2004).

We will evaluate these criticisms throughout the remaining chapters of the book while examining the connection of globalization to such issues as immigrant communities, world-city projects and non-Western world cities.

Command and control centers of the global economy

In the past decade or so, world-cities research has gained near hegemonic status in urban studies. Not all scholars agree on the narrow focus and categorization of world-cities research, but few dispute the importance of a global perspective in understanding urban economic change in the contemporary world. This section investigates those features of leading world cities that enable them to command the world economy – the characteristics of world cities that fascinate many urban scholars.

The conceptualization of global cities as the command and control centers of a globalizing economy draws greatly on Sassen's works (1991, 1994 and 2002). In her explanation of the emergence of global cities, Sassen points to the opposing geographical trends of dispersal and centralization that accompany globalization. The global dispersal of economic activities, helped by both space-shrinking technologies and deregulation measures, creates a huge demand for expanded central management functions. These functions include corporate headquarters and advanced business services, such as accounting, advertising, consulting, and financial and legal services (Table 5.4). These services, according to Sassen, tend to concentrate disproportionately in large global cities, such as London, New York and Tokyo, where their operation can benefit from "territorialized business networks" – in other terms, institutional thickness (Thrift, 1994) and territorial embeddedness (Budd, 1999).

It may still be arguable whether global cities are truly "a new type of city" in the age of globalization (Sassen, 1991: 4) and whether their emergence marks "a qualitatively new phase in urban development" (Taylor, 2004: 27). Regardless, the general consensus is that global cities are the key locations of most aspects of globalization. The majority also agrees that the continuing globalization of finance, and its demand for centralized management and services, serves as a major cause of economic, cultural and spatial restructuring in global cities. An

Table 5.4 Top ten cities in Global 500

Rank	City	Number of Global 500 companies	Global 500 revenues ($ millions)
1	Tokyo	52	1,662,496
2	Paris	27	1,188,819
3	New York	24	1,040,959
4	London	23	1,054,734
5	Beijing	15	520,490
6	Seoul	9	344,894
7	Toronto	8	154,836
8	Madrid	7	232,714
8	Zurich	7	308,466
9	Houston	6	326,700
9	Osaka	6	180,588
9	Munich	6	375,860
9	Atlanta	6	202,706
10	Rome	5	210,303
10	Düsseldorf	5	225,803

Sources: CNNMoney.com (2006)

understanding among urban scholars that globalization and urban change in global cities are mutually constitutive, instead of the one being a mere outcome of the other, has spurred the rise of globalization–urbanization nexus literature in urban studies in the past decade.

Large numbers of corporate headquarters (Short and Kim, 1999), business service firms (Martin, 1999; Taylor *et al.*, 2002) and high-paid professional jobs (Florida, 2005a; Hartley, 2005) affirm the command and control functions of emergent global cities. To explain the spatial clustering of business services, many also point to the place-specific social and cultural determinants of global cities, such as the personal and professional networks of leading financial experts (Budd, 1999; Thrift, 1994). Furthermore, along with managing and servicing the global economy, global cities foster cosmopolitan cultures, which represent a significant social aspect of globalization (Abrahamson, 2004; Hannerz, 1996; Yeoh and Chang, 2001). According to Hannerz (1996), global cities (such as New York, London, Paris, Los Angeles and Miami) have built the status of the global cultural market place based on the presence – transitory and permanent – of transnational business elites, immigrants from developing countries, creative specialists and international tourists.

Some identifiable commonalities of global cities range from clustered financial firms through high rents in central business districts to growing social polarization among urban residents. Yet these cities also exhibit pronounced differences. For example, Tokyo's world-city status is often attributed to economic prowess, backed by the worldwide success of Japanese multinational corporations. Meanwhile, both London and New York have been able to complement their economic power with a long history of cosmopolitanism. New York is situated in a relatively decentralized national urban hierarchy, whereas London and Tokyo each exist in highly centralized ones. London, owing much to a dynamic process of regional integration in Europe, tends to rank higher than the other two in terms of international connectivity (Table 5.5). More comparative studies could look into issues of ethnic communities and politics of globalization in these cities.

The global urban hierarchy

The three leading world cities, London, New York, and Tokyo, stand in a league of their own. World-cities researchers often base the world-city status of other cities on how well, or how poorly, they measure up against these top three. In contrast to identifying common attributes of the three leading world cities, another major trend in world-cities research focuses on interconnections and hierarchical relations among major cities around the world. Heralded by Friedmann (1986), many have attempted to draw a global hierarchy of cities that

Table 5.5 International passenger traffic, April 2005–March 2006*

Rank	City (airport code)	Total international passengers**
1	London Heathrow (LHR)	61,020,878
2	Paris (CDG)	49,054,367
3	Frankfurt (FRA)	44,827,705
4	Amsterdam (AMS)	44,072,756
5	Hong Kong (HKG)	40,283,000
6	Singapore (SIN)	31,019,193
7	London Gatwick (LGW)	28,783,016
8	Bangkok (BKK)	27,301,098
9	Tokyo (NRT)	27,104,631
10	Seoul (ICN)	25,691,061
11	Dubai (DXB)	24,265,207
12	Madrid (MAD)	22,365,635
13	Munich (MUC)	19,661,594
14	London Stansted (STN)	19,431,361
15	Taipei (TPE)	19,419,425
16	Manchester (MAN)	18,678,463
17	New York (JFK)	18,518,876
18	Copenhagen (CPH)	18,250,947
19	Dublin (DUB)	17,841,216
20	Los Angeles (LAX)	17,481,706

Notes: * Airports participating in ACI's Monthly Traffic Statistics Collection
** International passengers: traffic performed between the designated airport and an airport in another country/territory
Source: Airports Council International (2006)

indicates the respective influence of individual cities on the current world economy (Alderson and Beckfield, 2004; Beaverstock et al., 1999; Keeling, 1995; Knox and Taylor, 1995; Short et al., 1996; Taylor, 2004). The hierarchical arrangement of world cities, "in accord with the economic power they command" (Friedmann, 1995: 25), naturally begs the question of indicators and data. However, the lack of data hampers efforts in measuring and comparing the economic command of individual cities across the globe. Along with the lack of comparable data at the city level, Taylor (2004: 31–42) notes that the absence of academic efforts to provide further specifications and evidences for the global urban hierarchy in question has caused the crisis of "theoretical sophistication and empirical poverty" in world-city literature.

In recent years, a great deal of scholarship has looked to data sources that could ensure a more empirically grounded, if not sound, hierarchy of world cities. Some have applied highly sophisticated statistical measures, such as network analytic techniques and blockmodeling techniques, to previously known data sources, including the Fortune Global 500 (Table 5.6). Many have focused on comparing one or two indicators, such as business services, multinational firms and/or international airline networks, to complete their hierarchy, instead of comparing exhaustive aspects of urban economic power. Some explore new indicators that could reveal the global influence of individual cities. These indicators include the hosting of international sports competitions (Short *et al.*, 1996), the concentration of global entertainment industries (Abrahamson, 2004), and the ratio of foreign-born residents to the total population (Benton-Short *et al.*, 2005) (Table 5.7). Some have attempted to devise a composite index that quantifies various world-city indicators (for example, Cai and Sit, 2003), yet without convincing results.

An investigation of "empirical poverty" has been led by the Globalization and World Cities (GaWC) Study Group and Network, created in 1998 by Peter Taylor and Jon Beaverstock at Loughborough University. GaWC has carried out extensive data collection and quantitative network analysis to establish intercity relations at the global scale (Taylor, 2004). In addition to creating a large data matrix, GaWC has gained a strong reputation for its core members' productivity and scholarship. Indeed, its website has emerged as the essential site for global-cities research, listing latest research outcomes and, consequently, controlling major issues and topics related to global cities in the past few years.

Table 5.6 The world's ten most "powerful and prestigious" cities*

Rank	Power			Prestige
	Outdegree centrality	Closeness centrality	Betweenness centrality	Indegree centrality
1	Tokyo	Paris	Paris	New York
2	New York	Tokyo	Tokyo	London
3	Paris	London	Düsseldorf	Paris
4	London	New York	London	Tokyo
5	Düsseldorf	San Francisco	New York	Los Angeles
6	Amsterdam	Düsseldorf	San Francisco	Chicago
7	Zurich	Amsterdam	Munich	Brussels
8	Munich	Munich	Oslo	Amsterdam
9	Osaka	Chicago	Vevey	Singapore
10	San Francisco	Stockholm	Zurich	Hong Kong

Note: * Calculated from the data on the headquarters and subsidiary locations of the Global 500 in 2000
Source: Alderson and Beckfield (2004: 830)

Table 5.7 World cities in a culturally globalizing world

Global cultural industries hierarchy (recorded music, movies and television)*

1st tier	New York
2nd tier	London, Los Angeles, Paris, Sydney, Tokyo
3rd tier	Toronto
4th tier	Cairo, Hong Kong, Luxembourg, Manila, Mexico City, Mumbai, Nashville, Rio de Janeiro
5th tier	Brussels, Miami, Montreal, Washington, DC

Urban immigration index cities**

1st tier	New York, Toronto, Dubai, Los Angeles, London, Sydney, Miami, Melbourne, Amsterdam, Vancouver
2nd tier	Riyadh, Geneva, Paris, Tel Aviv, Montreal, Washington, DC, The Hague, Kiev, San Francisco, Perth
3rd tier	Munich, Calgary, Jerusalem, Boston, Chicago, Ottawa, Edmonton, Frankfurt, Winnipeg, Brussels, Düsseldorf, Seattle, Rotterdam, Houston, Brisbane, San Diego, Copenhagen, Bonn, Detroit, Milan, Cologne, Zurich, Rome, Berlin, Vienna, Portland, Hamburg, Minneapolis–St. Paul, Singapore, Stockholm, Dallas–Fort Worth, Tbilisi, Quebec City, Buenos Aires, Oslo

Sources: *Abrahamson (2004: 159); ** Benton-Short et al. (2005: 955)

Although GaWC's website covers almost all aspects of global cities and their network, its core members pay particular attention to the role of advanced business services both in globalizing the current world economy and in connecting major cities around the world. One emerging database, named the GaWC 100, is based on headquarters and subsidiary locations of 100 leading global service firms, including accountancy, advertising, banking and finance, insurance, and law and management consulting firms. The raw data are used to calculate total service value and global network connectivity for 315 cities (Figure 5.2). Based on this data matrix, Taylor (2004) establishes a "world-city network", in which global cities function as nodes in transnational economic flows.

It is still too early thoroughly and fairly to evaluate GaWC's contribution to our understanding of global cities and, in broader terms, the globalization and urbanization nexus, since a large number of its projects are still ongoing. While his data collection and network analysis are innovative, Taylor's (2004) maps and configurations demonstrating urban global network connections do not match the sophistication of his theoretical argument for intercity relations at the global level. Some also claim that GaWC pays little attention to numerous large cities that do not represent a market for leading service firms – Short (2004a: 50) calls them "black holes" of advanced global capitalism. Hall and Pain (2006) urge GaWC to

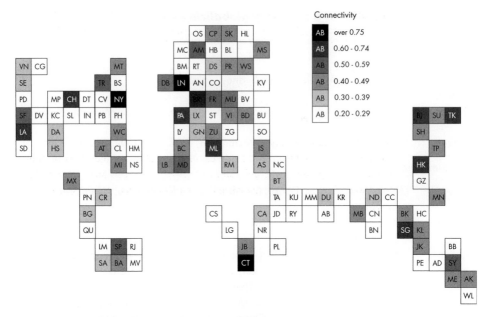

Figure 5.2 GaWC's global network connectivity

Note: For city codes, see Taylor (2004: 72, Figure 4.1)
Source: Taylor (2004: 73)

compare "global urban regions," not global cities, since a long-continued process of decentralization has produced many mega-city regions (e.g., Southeast England and the Paris region) that now function as nodes in the global urban network. Regardless of its shortcomings, GaWC has heightened the importance of establishing empirical, quantitative bases for the discussion of global urban hierarchy, in which lists of global cities were often presented without minimum specifications.

Further reading

Friedmann, John, 1986, "The world city hypothesis," *Development and Change*, 17, pp. 69–83. This is a seminal article on world cities and their vital role in the globalizing world economy. It should be a starting point for anybody who attempts to understand world cities as both a concept and a phenomenon in the era of globalization.

Knox, Paul L. and Peter J. Taylor, eds, 1995, *World Cities in a World System*, Cambridge: Cambridge University Press. An important collection of essays by leading scholars who explore theoretical and empirical links between world-cities research and world-system analysis. Part I of the book provides an excellent overview of conceptual and theoretical frameworks that are used in world-cities research.

Sassen, Saskia, 1991, *The Global City: New York, London, Tokyo*, Princeton: Princeton University Press. In this classic book on the three power centers of the world economy, Sassen conceptualizes global cities as the command and control centers of the global economy.

Taylor, Peter J., 2004, *World City Network: A Global Urban Analysis*, London: Routledge. Criticizing the existing studies on world cities for their lack of both theoretical interrogation and empirical evidence, Taylor illustrates in detail the specification of a world-city network and data collection and creation to measure it.

6 Globalization and globalizing cities

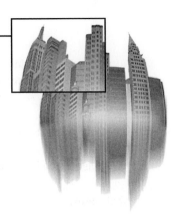

Learning objectives

● To think about what makes a city competitive in a globalizing world economy
● To gain insight into the politics of world-city status
● To understand the interplay between globalizing and localizing forces in urban change
● To explore and compare different world-class city claims made by second-tier cities

The urban impact of globalization is not limited to leading world cities. Other cities, though not as economically powerful as the top-tier world cities, also undergo various forms of economic restructuring in relation to the forces of globalization. The nature and extent of urban changes in these non-world cities warrants attention, yet many world-cities researchers have assumed, consciously or unconsciously, that smaller, more ordinary cities experience a lesser version of the same kind of urban restructuring observed in world cities. This assumption overlooks vast differences in the urban experience worldwide. The diversity of urban dynamics prompts the need for case studies representing a range of population sizes, national economies, cultures and political systems.

World-cities researchers more or less agree on the economic characteristics that distinguish leading world cities from non-world cities. Yet, the shared characteristics of London, New York and Tokyo do not necessarily translate into what it takes to be a world city. Few cities around the world can even dream of housing the globally commanding stock markets, corporate headquarters, bankers, lawyers and real estate developers that have made the top three so special. Benefiting naturally from reputation, governments of the leading world cities see little need to advertise their assets to potential businesses, residents and tourists. However, that is not the case for the second-, third- or lower-tier world cities. Even those well positioned in their domestic setting might find themselves fairly anonymous among international audiences. Promoting cities in international markets presents a challenge to urban governments and, in some cases, to national governments. Still, some promotional strategies have proved effective to varying degrees. This chapter reviews such efforts and challenges faced by cities striving for a more global status.

To understand better the globalizing processes of non-world cities, certain questions must be asked, including:

- What makes a city competitive in an increasingly globalizing economy?
- How do urban economic development strategies rework the forces of globalization?
- What policy initiatives promote world-city status or urban success?

Before we attempt to answer these questions, let us think about how much policy decisions account for change in urban fortune over time. While urban politicians generally take credit for any positive economic change in their cities, they are quick to blame negative change on structural constraints. Either way, they contend that government policies for urban development are more relevant than ever for city competitors in the uncertain game of globalization. Ever more politicians use terms like "urban success," "urban competitiveness" and "world-class city" in their vision statements, though often without much justification. In the recent rise of urban entrepreneurialism, such scripted messages of global competitiveness have replaced many traditional agendas, such as public services and housing projects. Some well-designed, well-executed urban policies may indeed improve the international status of a city, although most policies fall short of this goal. City governments, regardless of a repeating failure pattern of policy outcomes to match policy expectations, continue to take on new initiatives for urban competitiveness. We are very interested in the politics behind such economic policies, because urban changes are often caused not so much by globalization as by globalization-related policies.

This chapter begins with a section on urban competitiveness, a concept that highlights the role of urban governments in promoting economic development within a globalizing world. The second section critically examines the policy initiatives that many large cities in the developed world have employed in pursuit of world-city status. The third and final section reviews the homogenized image of globalized urbanism and the academic endeavor to recognize the diversity of urban experiences in a global capitalist system.

Urban competitiveness

Economists, for all their expertise, offer no guaranteed strategies for urban economic development (O'Flaherty, 2005). Nor can other people, they insist. Yet, most city officials throughout the developed world, particularly American city mayors, seem sure that they know what it takes to achieve economic development (Box 6.1). Often such leaders project an image of impending, unprecedented economic success and world-class city status. There is, however, a marked gap between the rhetoric and the reality of urban economic development. This section examines the nature of urban economic development and the policies and programs proposed to achieve it, while the next section examines the policy initiatives specifically related to world-city status.

In today's globalizing world, ever more cities are developing close economic relations with cities beyond their national boundaries. In operating their global production and marketing networks, multinational firms connect various international cities that otherwise would rarely interact. New direct international flights generate an array of new transactions among the cities involved. In this sense, it seems more accurate to define globalization as the rise of intercity activities, rather than international activities.

Urban policies on economic development are increasingly framed in the context of urban competitiveness in international markets (Kresl and Fry, 2005; Savitch and Kantor, 2002). Many traditional industrial cities in North America and Western Europe, which have suffered deindustrialization and decentralization for decades, aggressively seek to reverse their urban fortunes by promoting their competitive advantage in international service industries. Meanwhile, newly industrializing cities seek investments beyond their national borders, capitalizing on a greater locational flexibility of manufacturing and service industries in the contemporary world.

How can cities, and especially their governments, compete in this market situation? Ardent proponents of the free market claim that city governments can do little for economic development, since the market tends to achieve higher productivity and efficiency when left alone. However, political action, along with geography, history and other local factors, represents an essential element in urban economic change within today's global capitalism. Here we identify two opposing arguments for how individual city governments should respond to the forces of globalization.

One could argue that, since economic development yields huge net external benefits for urban residents and businesses, governments have a responsibility to facilitate such development. Accordingly, business incentives and other policies

Box 6.1 American city mayors' commitment to economic development

Akron, Ohio

The Mayor's Office of Economic Development serves as the direct liaison for the City of Akron to the business community. Our goal is to create attractive opportunities for existing Akron companies to successfully grow their businesses and to assist companies and entrepreneurs looking to start-up or relocate new business operations in the Akron area. To meet this goal, our staff philosophy is to think outside the box and provide businesses with creative economic development opportunities, alternatives, solutions, and assistance to meet their specific needs and goals.

Atlanta, Georgia

Cooperation between business and government is an integral component of a world class city. Accordingly, Atlanta's administration is dedicated to cultivating that relationship to the benefit of all its citizens. The Mayor has taken several steps to prioritize economic development and help the city grow. She has introduced a coordinated multi-pronged effort to increase Atlanta's ability to attract, retain, and grow businesses and civic organizations while providing related infrastructure services. The administration is also working to increase the city tax base for the funding of necessary city operations and infrastructure.

Buffalo, New York

From the birth of our city in the early 19th Century through the remarkable growth period of the 20th Century and now in an era of change and rebirth during the 21st Century, Buffalo is prepared to build upon the many assets that makes us unique and attractive for investment, economic development and growth. Like many of our peer cities in the Northeast, our city has weathered challenging economic times, yet we have remained steadfast in our determination to move progressively and positively toward a better future. As Mayor, I am committed to harnessing the physical, cultural and intellectual strengths of our city and transforming those matchless qualities into sustainable and unrivaled developments that will benefit our citizens and visitors alike.

Sources: Official city websites

Case study 6.1

Urban decline in Ohio

"Urban decline" is a familiar term in the Western world, particularly within the Anglo-American urban experience. Viewed from a historical perspective on urban development, the decline of industrial cities might be an inevitable outcome of economic transformation, namely deindustrialization and a global shift in manufacturing. Yet, that is only half of the story. Urban decline in the Northeast of the US has been accelerated and magnified by social and political processes, including the suburbanization of white middle-class households (popularly termed "white flight"), political fragmentation and localism within a metropolitan area, and immigration, as well as community failure in inner-city neighborhoods.

Urban decline has been analyzed and portrayed in various ways. Population change provides a simple but effective method to examine the extent of urban decline. Table 6.1 shows population trends in seven Ohio cities during the twentieth century. All of them, except Columbus, experienced severe population losses in the second half of the century. They have continued to decline in population since the turn of the new century. Indeed, Cincinnati has been named, by the US Census Bureau, the number-one city in terms of population loss: its population dropped 6.8 percent in five years (from 331,285 in 2000 to 308,728 in 2005). Its in-state rival Cleveland hardly fares better. The city's current population amounts to less than half of its all-time high in 1950. Cleveland's economic crisis appeared on national media in 2000 when the city ranked first for highest percentage of people below the poverty level: 31.3 percent. It is therefore ironic that Cleveland was selected as one of the eight most successful US cities in a survey of urban experts on the cities "that are perceived to have experienced the greatest economic turnaround or revitalization" between 1990 and 2000 (Wolman et al., 2004).

Table 6.1 shows how the rankings of seven Ohio cities have fluctuated in the list of America's 100 largest cities in the past two centuries. Cincinnati was one of the top ten largest cities in the nation for most of the nineteenth century. Cleveland made the top ten later, and maintained that status until the 1970s. Since then, both cities have slipped significantly in the rankings. The national rankings for the four medium-sized industrial cities in the state – Akron, Dayton, Toledo and Youngstown – peaked in the first half of the twentieth century, during which time Ohio boasted seven cities ranked within the top fifty. However, times have changed. Today, few residents in those four cities would guess that their respective cities once ranked among the nation's largest.

Population losses in the central cities have swept across the Northeast and Midwest, but Columbus has managed to avoid this regional trend. The city gained population

Table 6.1 Population decline in large industrial cities in Ohio, 1900–2000

	Akron	Cincinnati	Cleveland	Columbus	Dayton	Toledo	Youngstown
1900	42,728	325,902	381,768	125,560	85,333	131,822	44,885
1910	69,067	363,591	560,663	181,511	116,577	168,497	79,066
1920	208,435	401,247	796,841	237,031	152,559	243,164	132,358
1930	255,040	451,160	900,429	290,564	200,982	290,718	170,002
1940	244,791	455,610	878,336	306,087	210,718	282,349	167,720
1950	274,605	503,998	914,808	375,901	243,872	303,616	168,330
1960	290,351	502,550	876,050	471,316	262,332	318,003	166,689
1970	275,425	452,524	750,903	539,677	243,601	383,818	139,788
1980	237,177	385,457	573,822	564,871	203,371	354,635	115,436
1990	223,019	364,040	505,616	632,910	182,044	332,943	95,732
2000	217,074	331,285	478,403	711,470	166,179	313,619	82,026

Source: US Census Bureau (various years)

Table 6.2 Ranks of industrial cities in Ohio by population of the 100 largest US urban places, 1810–2000

Year	Akron	Cincinnati	Cleveland	Columbus	Dayton	Toledo	Youngstown
1810	–	46	–	–	–	–	–
1820	–	14	–	–	–	–	–
1830	–	8	–	–	–	–	–
1840	–	6	–	–	–	–	–
1850	–	6	41	37	61	–	–
1860	–	7	21	49	45	68	–
1870	–	8	15	42	44	40	–
1880	–	8	11	33	47	35	–
1890	–	9	10	30	45	34	91
1900	87	10	7	28	45	26	84
1910	81	13	6	29	43	30	67
1920	32	16	5	28	43	26	50
1930	35	17	6	28	41	27	45
1940	38	17	6	26	40	34	49
1950	39	18	7	28	44	36	57
1960	45	21	8	28	49	39	75
1970	52	29	10	21	59	34	98
1980	59	32	18	19	70	40	–
1990	71	45	23	16	89	49	–
2000	86	56	35	15	–	58	–

Source: US Census Bureau (1998 and 2006)

throughout its modern history, and, accordingly, its national rank has improved over the years. But why? The fact that the city's economy never depended heavily on manufacturing helped. The state government has created many jobs for city residents. Furthermore, Columbus continues to add new economic activities, such as medical services and financial services, to the old. Still, Columbus suffers many of the inner-city economic crises common to most large American cities. Powell (2004) points to its aggressive annexation policies as a major contributor to its steady growth in population over a long period of time. Columbus has expanded geographically through a series of annexations since the 1950s, and it has fewer suburbs than most other large cities in the US. Attracting new and large-scale private investment can help to cure urban problems in declining industrial cities, as can improved metropolitan governance, or, more realistically, less metropolitan fragmentation in urban governance (Bradbury et al., 1982).

to encourage private investment should deserve high-priority status in city government. According to economic development advocates, advancing urban competitiveness is one of the most important – if not the only – criteria for good governance in the era of globalization. As such, city mayors should have a strategic plan for global competitive development (Kresl and Fry, 2005).

Others argue that good governance should be measured instead by the provision of collective goods and public amenities, since no evidence indicates clearly that the benefits of economic development will necessarily reach the majority of city residents. Through this social development perspective, the notion of intercity competition appears as an exaggeration created by assorted political and business interests – groups that would reap the majority of development benefits. Furthermore, supporters of this view believe that, in the era of globalization, city governments should focus less on tax breaks and other business-friendly policies, and more on long-term public service provisions that mitigate the harmful effects of market forces on local residents and businesses.

Most city policy agendas combine elements of both economic growth and social development arguments into overlapping programs for workforce development, job creation and community development. Some cities may be constrained by inherent advantages or disadvantages relating to such factors as geographical location and political traditions. In addition, local governments may claim different levels of authority over their urban development, depending on their regional and national governmental structure. In their comparative study of ten cities across North America and Western Europe, Savitch and Kantor (2002) identify such policy-making variables as market conditions, intergovernmental support, popular control systems and local culture. Yet, they stress that in all ten cities, political leaders made active policy decisions for urban development – playing the cards they were dealt – resulting in remarkable variations among them.

Overall, economic development receives high-priority status from politicians in large cities in the developed world, and public spending on economic development incentives continues to expand (Gottdiener and Pickvance, 1991; Hall and Hubbard, 1998; Logan and Molotch, 1987). Politicians often build campaigns around the positive message of economic growth and national and global competitiveness. When delivered in a simple manner, such a message can prove an effective tool to gain public support. Basically, the message goes like this: "With a well-defined vision and well-executed policies, a city can develop solid infrastructure, a quality workforce and favorable business environments, thus attracting new businesses and residents to the city." Often the pride – however maneuvered and misguided – that accompanies rhetoric of "the nation's best," "twenty-first century" and "world-class city" effectively charms voters.

In addition to the market situations and political interests that foster urban competition, assorted news media constantly evaluate the competitiveness of cities based on such broad, and somewhat vague, indicators as attractiveness to private investment and quality of life for a highly educated workforce. Many question the validity of those rating systems, since their assessment criteria for "best practices" and "urban success stories" often appear arbitrary. Yet, that does not stop cities from bragging about their high media ratings as a business-friendly, safe and/or entertaining place. One such example is Denver, Colorado, whose official website boasts various magazines' ratings that extol the city's good and business-friendly governance:

> For the past five years, Denver has graced FORTUNE magazine's list of "Best Cities for Business." Each year since 2001, we've been one of Dun and Bradstreet's "Top 10 Cities for Small Business," and this year, TIME magazine named Mayor John Hickenlooper one of the "Top Five Mayors in the Nation." With our shining local heroes, national acclaim and international reach, you can feel the exciting rush of energy constantly brewing in the City.
> (Official website of City of Denver)

Which policies do city governments implement in order to strengthen their competitiveness and gain such recognition? O'Flaherty (2005) lists several prominent policy initiatives designed to promote urban economic development and, more specifically, to make a city desirable for potential investors. They include:

- lowering taxes;
- passing targeted tax cuts;

- offering tailored incentives, such as subsidies through infrastructure, land or training programs;
- linking specific incentives to specific outcomes;
- reducing regulations;
- allocating urban enterprise and empowerment zones for economically distressed inner-city neighborhoods;
- building new sports stadiums and arenas.

Many doubt the growth potential of these incentives. In particular, critics question policies for attracting investment to a particular city, as so many cities engineer fairly similar strategies and offer almost identical incentive packages. Another lingering concern surrounds whether new businesses, after receiving tax breaks and other deals, make any significant contribution to the city's tax revenues. Peters and Fisher (2004) sum up various concerns over business incentives with the following three questions:

1   Do business incentives cause cities to grow more rapidly than they would have otherwise?
2   If so, is the growth targeted so as to provide net gains to poorer communities or poorer people, or is it merely a zero-sum game?
3   How costly to government is the provision of these incentives compared to alternative policies?

Encountering these challenges to their business-friendly policies, city politicians often talk about the "psychological edge" that large-scale investments from high-profile companies, such as the Fortune 500, would bring to the city's residents and small businesses. The construction of new stadiums for professional sports teams has been justified in similarly social and cultural terms, such as the importance of generating civic pride. It is not uncommon to see a new policy agenda, originally advocated for economic reasons, later justified on non-economic grounds.

Politics of world-city status

Much like the concept of competitiveness, the term "world city" is widely used in public and political rhetoric. Many urban scholars seek to examine how urban politicians benefit from creating the vision of their city becoming a world-class city. Despite its frequent use, much dissension surrounds what constitutes a world city and, furthermore, how non-world cities can achieve world-city status. Indeed, the vagueness of this concept allows city officials and related interest groups freely to pursue the coveted status and to promote a set of strategies and initiatives that might benefit themselves more than their city's quest for that status. Politicians can strategically select images of a world city, then craft, articulate and subsequently promote that image to serve politically motivated agendas (Machimura, 1998; Paul, 2004).

To examine the politics of economic development, much of the new urban politics literature highlights the political and business interests behind urban boosterism, machine politics and entrepreneurial governance in the US and beyond (Hall and Hubbard, 1998; Logan and Molotch, 1987). The notion of being or becoming a world city adds another dimension to these studies, and to existing debates on successful, desirable urban governance.

Three trends stand out in the rhetoric of world-city status. First, urban politicians aggressively promote the vision of their city as world class when they solicit public support for controversial plans, such as urban redevelopment projects, large-scale infrastructural upgrading and hosting or bidding for high-profile international events. Although debatable, the idea that such projects or events would bring world-city status to the city in question proves worth exploiting when the public responds favorably to the promise of international recognition for their city.

Second, politicians and business associations also utilize world-city rhetoric when arguing for large-scale tax incentives and business-friendly economic policies. While attempting to fulfill their ambitions, such politicians often associate the city's world-city status with its homegrown companies' global reputation and/or its ability to attract internationally famous businesses. The official websites of almost all US cities list the Fortune 500 companies that operate locally and, in some cases, their executives' testimonies of reduced tax rates and other governmental supports that have been offered to them.

Third, the world-city campaign is used to support attempts to heighten multiculturalism and, in a broader sense, cosmopolitanism among city residents. The imagineering of a world city that nurtures cosmopolitan cultures helps city governments solicit public support for costly cultural projects, such as ethnic festivals or the construction or renovation of opera houses and cultural centers. Cosmopolitan cultures often sustain successful tourist industries – one of the target sectors, along with high-tech industries, that almost all cities include in their urban economic development initiatives (Plate 6.1).

Despite all the rhetoric, it remains unclear whether world-city status can be achieved through a well-designed economic plan. The fact that the leading status of the world cities, such as New York and London, can hardly be linked to any government projects casts doubt on the effectiveness of world-city projects in non-world cities. Such reservations, however, have not stopped city governments from seeking world-city status.

Plate 6.1   Glasgow: Scotland with Style campaign

Since 2005 Atlanta has undertaken a comprehensive economic development strategy, named the "New Century Economic Development Plan for the City of Atlanta," to become the thriving core of the metropolitan area, the most successful city in the Southeast and a competitive city both nationally and internationally (Figure 6.1). After a lengthy analysis of the current state of its economy and the benchmarking of other cities, such as Seattle and Boston, the Atlanta city government has decided to focus on three economic priorities for achieving this grand vision: economic growth, physical infrastructure and safe, healthy neighborhoods (Table 6.3). According to the Development Plan, these priorities will be achieved through ten action plans, which involve countless "action items," "action owners" and "active partners." Ever more city governments produce their development plan in consultation with professional management firms, and Atlanta is no exception to that trend. Its official website notes that the Development Plan was produced with pro bono assistance from Bain & Company, a Boston-based global business consulting firm.

While some non-world cities draft equally impressive economic development plans for international competitiveness, some simply claim that they are already

Figure 6.1 Atlanta toward a world-class city

Source: **Official website of the City of Atlanta**

a world-class city and repeat the claim as often as possible, comparing themselves to emerging world cities or highlighting their similarity with the top three world cities. For example:

> San Francisco's vibrant business environment has more than 60,000 businesses and employs more than one-fourth of the region's total workforce. We sit at the center of global business activities, being located midway between London and Tokyo. Seventy percent of our residents receive training beyond high school, creating a highly educated and motivated labor force. We are a world-class city with a friendly neighborhood feel. Our diverse population enhances our everyday life and enriches our business community. In fact, experts say our diversity makes us one of the most productive areas in the country.
>
> (Official website of the City of San Francisco)

City mayors are not alone in exploiting the vision of their cities becoming legitimate world cities. Local and national news media act as another primary agent, not only in transmitting selected images of a world city, but in developing

Table 6.3 Atlanta's New Century Economic Development Plan

Economic Vision	Atlanta will be the thriving core of the metropolitan area, the most successful city in the Southeast, and a competitive city, nationally and internationally

Economic Priorities	• Economic opportunity:
	    job creation
	    business climate
	• Physical infrastructure
	• Healthy neighborhoods and quality of life:
	    workforce housing
	    public schools
	    public safety
	    increase economic vitality of underserved areas
	    parks and green space

Action Plans	• Support growth of target industries • Create and grow business recruitment, retention, and expansion • Champion BeltLine, Downtown and Brand Atlanta Campaign as major development projects • Increase economic vitality in underserved areas • Make it easier to develop in Atlanta (business climate) • Increase workforce housing • Increase capital available for development and business growth • Make Atlanta one of America's safest cities (crime rate) • Collaborate to improve the graduation rates in Atlanta public schools • Grow dedicated parks and greenspace

Source: Official website of the City of Atlanta

those images. They often fuel their city's hyped rivalry with other cities world-wide, as demonstrated every four years during the bidding for the Olympic Games (Short, 2004a). Local media often criticize individual politicians when their city fails to attract international sports events. They also help to promote highly selective, often misleading images of a world-class or Olympic city, as well as the "best practices" for becoming one.

Globalizing cities and globalized urbanism

Globalization does not mean the homogenization of the world. The notion of a homogenizing world goes back to early globalizationists like Kennich Ohmae, who linked globalization to the "universalizing" power of certain economic, political and cultural forces across national boundaries. However, the suggestion that distinctive, place-bound politics, traditions and practices at the local level could be erased completely by homogenizing forces of globalization is neither empirically supported nor theoretically sound. Instead, many insist on the continuing significance of government policies, cultural differences and national identities, even as they interact with global forces. None of the previous historical processes, such as modernization, industrialization, capitalism and colonialism, changed the whole world in such a similar manner and to such a similar extent as does globalization. However, those processes were as revolutionary in their day as globalization is in ours. It is fair to assume that a truly globalized world will be anything but homogeneous.

Still, a strong trend of convergence can be observed among various cities around the world. Different downtown landscapes, filled with almost identical office buildings, easily recognizable retailers and common fashion trends, greatly resemble each other. Globalized consumption patterns have materialized in large cities in both the Western and the non-Western world. This phenomenon has captured the attention of travelers and journalists as well as academics, most of whom attribute it to the global imperatives of late capitalism, postmodern capitalism and capitalistic consumer culture. In other words, they argue that different parts of the world have gradually integrated into one capitalist system. Certainly, globalized consumption patterns make many different places look, feel and taste the same. Notions of global urbanism or globalized urbanization highlight considerable common ground among cities with different traditions and political settings. One might argue that, as these cities increasingly adopt similar architectural styles, business-friendly policies and cosmopolitan lifestyles, they will eventually yield carbon copies of the same capitalistic, postmodern, Western city.

Compelling as the globalized urbanism argument sounds, it draws greatly either on sketchy observations of commonly visited sites around the world or on world cities and "wannabe" world cities. Viewed from above or from their downtown business districts, many cities share common characteristics of global development, but "looking closely on the ground," one observes distinctly local activity alongside those globally recognized practices and structures (Jordan, 2003: 32). Meanwhile, the tendency of world-cities research to focus on a handful of large cities overlooks both smaller cities in developed countries and almost all cities in developing countries (Gugler, 2004; Krause and Petro, 2003; Öncü and Weyland, 1997). The neglect of so-called non-global cities or ordinary cities limits "our imaginations about the futures of cities" (Robinson, 2002: 535).

A new wave in urban studies looks beyond the category of world cities to smaller cities. Such efforts observe whether smaller cities integrate into the global economy or become marginalized by it; whether they radically change with, gradually adapt to, or resist global trends; and whether they benefit or suffer from such processes. Although it does not ensure due academic attention to all deserving cities across the globe, this new trend in world-cities literature has encompassed a range of cities from Lexington, Kentucky, (McCann, 2004), to Moscow (Barter, 2004) to Accra (Grant, 2002).

This research expansion has also prompted a meaningful shift in the focus of globalization–urbanization literature from hierarchical categorization to dynamic urban processes. While early studies focused more on identifying aspects of emergent world cities, many now look to delicate changes in ethnic neighborhoods and architectural designs, as well as to rhetorical trends in urban politics pursuing world-city status. These micro-scale studies note the coexistence of global space (e.g., Wal-Mart stores) and local space (e.g., public buildings with historical memory), and attempt to uncover the dynamics of globality and locality in detail (Jordan, 2003). Individual places, they find, incorporate and rework globalizing effects, instead of being completely consumed by them. Smith (2001) presents cities as "translocalities" filled with social and political actors who localize national and transnational practices. He continues:

> A further advantage of this approach to urban studies is that a wide variety of cities, rather than a handful of sites of producer service functions, or a score of interwoven, mainly Western, centers of global command and control, become appropriate sites for comparative research. Viewing cities as contested meeting grounds of transnational urbanism invites their comparative analysis as sites of: (a) the localization of transnational economic, sociocultural, and political flows; (b) the transnationalization of local socioeconomic, political, and cultural forces; and (c) the practices of the networks of social action connecting these flows and forces in time and space. These emergent transnational cities are human creations best understood as sites of multicentered, if not decentered, agency, in overlapping untidiness. This is a project in which studying mediated differences in the patterns of intersecting global, transnational, national, and local flows and practices, in particular cities, becomes more important than cataloging the economic similarities of hierarchically organized financial, economic, or ideological command and control centers viewed as constructions of a single agent – multinational capital.
>
> (Smith, 2001: 70)

Smith (2001: 175) notes that a close look at localized practices of transnational connections should not be seen as "privileging local knowledge or essentializing the local community as a sacred space of ontological meaning." Instead, he calls for more comparative urban studies on the ways in which different transnational networks operate in the same city, as well as in which the same transnational network practice operates across different nations and cities.

The fact that urban places undergo change in different but comparable ways promotes comparative studies as well as individual case studies at various sites and under various circumstances (Grant and Nijman, 2002). Cities, no matter how localized their cultures and practices, exist as distinct units across a continuous, global grid, rather than as isolated, exceptional spheres (Hannerz, 1996; Oren, 2003). Their actions on, and reactions to, global forces can be compared, not only in terms of individual distinctiveness and locality, but in terms of commonality and globality. The interface of globalization and urbanization in globalizing cities could generate a large body of comparative studies, based on thorough case studies of various urban changes caused by, and causing, globalization.

Further reading

Hall, Tim and Phil Hubbard, eds, 1998, *The Entrepreneurial City: Geographies of Politics, Regime and Representation*, Chichester: John Wiley & Sons. This edited collection pulls together a range of detailed chapters on the recent rise of urban competition and entrepreneurial politics in Western cities.

Logan, John R. and Harvey L. Molotch, 1987, *Urban Fortunes: The Political Economy of Place*, Berkeley: University of California Press. This is a classic book on the politics of local economic development in American cities. It examines how American cities have become growth machines in which the growth ethic pervades virtually every cultural, economic and political institution that operates on the urban scene.

Paul, Darel E., 2004, "World cities as hegemonic projects: the politics of global imagineering in Montreal," *Political Geography*, 23 (5), pp. 571–96. This article is especially useful in studies of the politics of world city status in so-called second-tier world cities, such as Montreal.

Smith, Michael P., 2001, *Transnational Urbanism: Locating Globalization*, Oxford: Blackwell. Criticizing the globalization literature for exaggerating the universalizing power of late capitalism, Smith argues for the concept of transnational urbanism, which would allow us to examine the localized specificities of transnational practices across different cities.

7 New solutions for old economies

Learning objectives

- To understand the changing relationship between economies and cultures
- To define the creative class and its role in the new urban economy
- To assess critically the effects of high-culture- or creativity-driven urban development strategies

In the previous two chapters, we examined economic change in large cities, both world cities and second-tier cities. Urban change in second-tier cities often involves intense government efforts in pursuit of urban competitiveness in the international market place. Their city halls readily draft and subsequently implement a host of development strategies and action plans that promise to restructure them into world-class cities. In doing so, urban politicians want to be recognized and appreciated for their forward-looking vision that accommodates change in a globalizing world.

We now have a set of questions about cities for which achieving world-city status is not an imminent, or even realistic, goal for economic development policy:

- What kind of status do their economic development offices pursue when attempting to boost urban competitiveness?
- How do these cities consider the notion of globalization while assessing their current and, more importantly, future state of economy?
- What economic activities do declining industrial cities and their mayors aspire to develop?

In an attempt to answer these questions, we acknowledge new voices in the urban economic development discourse which call for creative strategies to tap into the so-called "new economy." The recent surge of public and political interest in the new economy does not necessarily propose an imminent sea change in the urban economy, as much as new ways to frame economic restructuring. The conceptualizations of and opinions about the new economy vary tremendously. Still, cities receive a great deal of advice on what policies can best reshape their old economy into a new one.

Smaller cities in general are neither endowed with a diversified economic base nor equipped with management skills to mitigate the effects of large-scale economic restructuring. Deindustrialization, one such restructuring, has devastated the livelihoods of millions in numerous traditionally industrial towns, even in previously thriving economies. Having suffered a sustained economic fall and its side-effects, the governments of many industrial cities turn to new sectors that promise to bring much-needed vitality to their local economies. These sectors include high-tech industries, tourism, finance, culture and the arts, among others. The promotion and development of such industries has become the policy of choice for many city governments, regardless of their political, geographical or historical conditions.

Out of countless new visions, new ideas and new solutions introduced to revitalize old, declining industrial cities, two key areas have received the most attention, both politically and academically: the cultural economy and the creative economy. This chapter first examines how these newly emerging economic activities relate to the revival of the declining urban economy and then discusses how politicians exploit these terms and the situations applying to them.

The cultural economy

City governments seeking solutions for their struggling economies increasingly turn to post-industrial and postmodern economic sources. Cultural activities, previously deemed to have marginal effects on the city's overall economic health, have won a newfound respect from policy-makers and, subsequently, increased public investment. City governments now consider funding cultural and arts institutions as an economic development measure. The rising status and potential of culture in the urban economy have both reflected and prompted scholarly investigations into the relationship between economics and culture in the capitalist system.

The "cultural turn" has been one of the most significant developments in critical social analysis in the past two decades, although cultural theories and studies started attracting attention as early as the 1960s. Simply put, a growing interest has developed in examining the ways in which cultural processes and institutions condition economic and political aspects of society. Indeed, the cultural turn has had a significant influence on the current intellectual generation across the social sciences, including some economists who have tried, explicitly or implicitly, to understand the social and cultural construction of economic activities (Granovetter, 1985; Throsby, 2001). The widely used business term "cultural embeddedness," for example, suggests a new thinking in corporate management that recognizes the value of non-economic factors, such as social networks and local practices, in achieving business goals, as well as in improving daily operations.

Along with the cultural context of economic activities, the economic dimensions of culture are a critical notion that defines the cultural turn in the social sciences. The fact that high-quality arts, antiques and collectibles carry enormous price tags illustrates the economics of certain cultural goods and services, though limited to a small faction of the more affluent and philanthropic society. Mass culture has also been commodified, mostly through tourism, consumer culture, media culture and youth culture. This worldwide commodification has been associated with late capitalism by both Marxists and postmodern theorists (Featherstone, 1995; Harvey, 1989b; Jameson, 1991; and Lash and Urry, 1994). Jameson (1991) notes that, in the latest stage of capitalism, symbolic meanings and associations increasingly determine the economic value of goods. In a postmodern consumer culture, according to Jameson, the strategy of imposing symbolism, meanings, values and emotions on to goods begins to blur the boundary between the image of those products and their concrete reality.

Upon recognizing the intrinsic links between economics and culture, the question becomes: how does culture translate into an urban economic sector? To answer this question, cultural activity must be defined as an industry that generates employment, tax revenues and multiplier effects. Throsby (2001: 111) describes this industrial aspect of creative works:

> [T]he fact that individuals and firms produce goods or services for sale or exchange, or even simply for their own pleasure, creates a grouping of activity around particular products, types of producers, locations, etc., which can be encircled in conceptual terms and labeled an industry . . . Nevertheless, it has to be conceded that in practice the application of the word "industry" to art and culture does focus attention on the economic processes by which cultural goods and services are made, marketed, distributed and sold to consumers. The term "cultural industry" in contemporary usage does indeed carry with it a sense of the economic potential of cultural production to generate output, employment and revenue and to satisfy the demands of consumers, whatever other nobler purpose may be served by the activities of artists and by the exercise of the tastes of connoisseurs. Indeed many within the cultural sector, including presumably those artists whose objective functions contain some

component of economic gain, welcome the idea that cultural activity makes a significant contribution to the economy.

Once it is agreed that culture contributes to the urban economy, the next debate revolves around what comprises the cultural industry. According to Throsby (2001: 112), the industry includes music, dance, theatre, literature, the visual arts, crafts and more recent practices such as video art, performance art, computer and multimedia art. Scott (2000) introduces a broader list of "cultural-products sectors," which includes, in addition to Throsby's artistic industries: high fashion, furniture, news media, jewelry, advertising and architecture. Sports industries, both professional sports and health and fitness-related businesses, could be added to the list (Dziembowska-Kowalska and Funck, 1999).

Scott laments that the US Standard Industrial Classification provides no explicit information on certain noteworthy segments of the cultural economy, including tourist services. Indeed, urban tourism ranks among leading activities of the cultural industry, as almost all cities now promote events and festivals in an effort both to generate tax revenues and to revitalize inner-city neighborhoods. For example, Harlem in New York, a prototype of disadvantaged and racially segregated American urban sites, has recently centered its community development strategy on promoting cultural and ethnic events that target both visitors and residents (Hoffman, 2003). The rapid growth of urban tourism proves that tourism comes not just to places blessed with historical treasures, exotic nature or other pronounced attractions, but to those that create the events, images, traditions and meanings that appeal to potential tourists (Plate 7.1).

Measuring the urban economic contribution of cultural activities is tricky. Totaling the revenue of large arts institutions and cultural events leads to a significant understatement of the cultural sector, since this does not account for the contribution of its multiplier effects in local businesses. Furthermore, insufficient data exists reliably to estimate these multiplier effects. However, Markusen and Schrock (2006), in their comparative and historical research of the twenty-nine US metropolitan areas with the largest presence of working artists, propose an even more comprehensive measurement, named the "artistic dividend" for a regional economy. They argue that the work of creative artists enhances the design, production and marketing of products and services in other sectors. All things considered, artistic and cultural activities induce innovation on many levels of the urban economy, which may not be easily measured in hard dollars.

Cultural industries concentrate in world cities and other large cities. Indeed, they contribute immensely to the economies of cities like Los Angeles and Paris (Power and Scott, 2004; Scott, 2000 and 2005). The concept of "path dependence" can be used to analyze cultural production as well as technological innovation

Plate 7.1   City for the Arts, Arts for the City, San Francisco, California

(Rantisi, 2004). Path-dependent theories claim that small historic events or locational advantages can affect macroeconomic consequences that privilege certain paths to development and limit others. Although criticized for its overly structural approach to local economic development, this idea sheds light on the prestigious success that Paris enjoys in high fashion, New York in advertising and Los Angeles in motion-picture entertainment. Their leading roles in these industries have long enjoyed wide recognition, and their advantages over potential competitors relate not only to the quality of their products but to the "symbolic images" – such as authenticity and reputation – that those products carry.

For the economic development of smaller cities, the notion of path dependence presents as many questions as answers. Do smaller cities ever profit from cultural industries? Can smaller cities cultivate local preconditions for a competitive cultural industry? Can cultural industries bring sizable employment, tax revenues and multiplier effects to other industries in medium-sized cities, even after their governments provide them with tax breaks or other subsidies? As noted in the previous two chapters, no one can yet guarantee the urban development potential of any policy measure, including the promotion of culture and the arts. This uncertainty, however, has not stopped city governments from adding cultural elements to their existing lists of development strategies (Box 7.1), or from commissioning studies on the potential economic benefits of new cultural projects. Most of these commissioned studies support the cultural project in question, citing likely benefits to local businesses and an increase in local property values. However, few cities commission studies to assess whether particular cultural projects have indeed delivered their promised benefits to local businesses and residents.

Culture and the arts have long played into urban redevelopment and gentrification (Neil Smith, 1996; Zukin, 1995). A great number of inner-city neighborhoods have

---

Box 7.1 Cultural development for traditional industrial cities

Glasgow, UK

Our vision for Glasgow: We want Glasgow to flourish as a modern, multicultural, metropolitan city of opportunity, achievement, culture and sporting excellence where citizens and businesses thrive and visitors are always welcomed.

Osaka, Japan

The role of Osaka and the Kansai region, as Japan's cultural and historic capital, will continue to grow in the future as the nation strives to meet the world's expectations for a greater contribution in the field of culture. The City of Osaka is making efforts to enhance its profile as an international cultural center by promoting cultural, artistic, academic, and sports activities in the city and enriching them through cultural exchanges with the rest of the world.

Toledo, USA

The mission of the City of Toledo's Department of Development is to improve the quality of life of our residents by coordinating city and other resources to create capital investment, job opportunities, safe and attractive neighborhoods, good-quality housing, and other important community services and facilities. This includes vibrant arts, cultural, and retail amenities for individuals and families.

Source: Official city websites

been "gentrified" with art museums, historic districts and, more controversially, professional sports stadiums. While the displacement of the existing communities has been detrimental to former residents, public funding for high culture often receives favorable reviews from local media and middle-class residents.

Medium-sized cities in Ohio favor renovating or expanding art museums to revitalize their city centers and, more importantly, to counter their troublesome industrial images. Toledo, for example, threw a grand opening ceremony in 2006 for the Glass Pavilion at the Toledo Museum of Art, which should remind both local residents and tourists of the city's once highly productive glass industry (now almost obsolete). As is shown in Box 7.1, the city's mayor sees cultural enrichment as part of his economic development mission. Cincinnati, another Ohio city suffering from a major manufacturing decline (particularly in aerospace and automotive industries), bought downtown property to build a new Contemporary Arts Center (Plate 7.2) in the late 1990s. This has since won the Pritzker Architecture Prize and become the signature building of its internationally renowned architect, Zaha Hadid (Sklair, 2005). However, brief fieldwork in the downtown area of Cincinnati in the fall of 2006 revealed that few city residents knew about their art center's fame, let alone its economic benefits to the city.

Plate 7.2 Contemporary Arts Center, Cincinnati, Ohio

Beyond the revitalization of inner-city neighborhoods, cultural strategies of urban development have focused on events, such as ethnic festivals, exhibits, performing arts and historical re-enactments, among others. City governments expect these events to attract tourists and suburban residents to their city center, providing indirect benefits to restaurants and hotel businesses in the downtown area. While such economic benefits may or may not materialize, local politicians certainly benefit from their upbeat presentation of culture, leisure, cosmopolitanism and postmodernism (Quilley, 2000).

The creative economy

Of countless tactics cities employ to compete in the twenty-first century, attracting creative talents seems a popular choice in both the media and political rhetoric. Meanwhile, as is often the case, many academics suspect the political logic behind this public policy phenomenon.

The notion of creative talent driving urban economic development draws almost entirely on Richard Florida's works (2002, 2005a and 2005b). Countless magazine articles and toolkits for urban programs contribute to the vision of the creative urban economy (Landry, 2000). Florida (2002) illustrates the emergence of a new social class, the creative class, which leads the new creative economy. Later, the same author (Florida, 2005b) warns that the American economy may be losing ground in global competition for creative talent. These two books define and promote the significance of the creative class in the national and regional economy, while *Cities and the Creative Class* (Florida, 2005a), outlines the geography of creative talents in the US, ranking cities according to a measurement of creativity, technology and diversity.

The concept of the creative class evokes both enormous excitement and harsh criticism in the urban economic development debate. The mainly positive media frenzy surrounding the term "creative talents" reflects a widespread belief in certain circles that a major economic overhaul is overdue. On the other hand, criticisms have emerged from many different factions in academia. Some claim that "informational economy" and "knowledge-based economy," which describe the roles of information processing and knowledge generation in capital accumulation, are more intellectually sound concepts (Castells, 2000; Dunning, 2000). These earlier terms have been used to suggest that new developments in information technology give rise to the capacity to generate and profit from new knowledge. Economically struggling cities, regions and nations have long discussed the benefit of developing industries relating to research and development and innovation. Indeed, many industrial cities in North America and Western Europe have attempted to develop such industries, but without much success. The notion of the creative economy might likewise inspire a wave of political and media policy discussions, without offering concrete opportunities for unemployed former factory workers. Below, we take a close look at Florida's thesis of the creative class.

Florida's creative capital theory states that "creative people power urban and regional economic growth, and these people prefer places that are innovative, diverse and tolerant" (Florida, 2005a: 34). Therefore, to succeed in today's world, cities must attract and retain highly creative people. According to Florida, the creative class divides into two groups of people. The first group makes up the "super-creative core," a broad and diverse group including scientists, engineers, university professors, poets, novelists, artists, entertainers, actors, designers and architects, as well as non-fiction writers, editors, cultural figures, think-tank researchers, analysts and other opinion-makers. The other group consists of "creative professionals" who work in a wide range of knowledge-intensive industries such as high-tech sectors, financial services, legal and healthcare professions and business management. Florida explains that the former group produces new forms or designs that are readily transferable and widely useful, while the latter group engages in creative problem-solving. He estimates that roughly 30 percent of the entire US workforce (about 38.3 million workers) belongs to the creative class.

Florida's concept of the creative class is not particularly new: human capital literature has long stressed the relationship between greater numbers of highly skilled, highly educated talent and urban growth (Glaeser *et al.*, 1992 and 1995). Jane Jacobs (1969) points to a steady stream of innovations, enabled by the geographical concentration of innovative people with diverse ideas and occupations, as a primary cause of the rapid growth of New York's garment industry – and the city's economy in general – in the nineteenth and early-twentieth century. Supported by empirical studies of American cities (Lucas, 1988; Romer, 1986), the notion that knowledge spillovers, spurred by thick interactions among a large number of highly skilled people living and working in geographical proximity, promote urban growth has become conventional wisdom in studies of new sources for economic development. Florida has expanded the existing notion of the human capital, comprising professionals in high-tech, research and advanced service industries, by adding occupations related to culture and the arts. He has branded these people as the "creative capital," a term that captures more media attention than the more bland "human capital."

Highly creative people indeed contribute to economic growth. However, other factors may explain the popularity of the creative class thesis in media and public policy (see the CreativeTampaBay in Box 7.2). Media and policy-makers may respond favorably to the suggestion of renovating urban cultures and aesthetics to create a more appealing environment for creative people (Plate 7.3). Florida proposes the development of a diverse urban culture as the most effective way to attract creative and highly skilled people, and thereby stimulate economic growth. To support this argument, Florida (2005a) compares selected American cities through creative measurements, including the coolness measure, the Bohemian index and the gay index, as well as numbers of college graduates and computer software industry workers. Not surprisingly, large coastal cities with quality research institutes and diverse cultures, such as San Jose, Seattle and Boston, rank high (Table 7.1). Florida is eager to point out that an open and tolerant culture is a primary cause of these cities' recent economic success.

Box 7.2   History of CreativeTampaBay, Florida

On April 11, 2003, almost five hundred people came to the Tampa Bay Performing Arts Center to hear creative economy guru Dr. Richard Florida describe what it takes to build a vibrant community in today's technology environment. It was music to the ears of most in attendance and the excitement and energy has been building ever since. Getting Dr. Florida to Tampa and attracting so many community leaders to hear him speak was the result of a coalition of organizations involved in business and the arts: The Greater Tampa Chamber of Commerce, Tampa Bay Partnership, the Pinellas and Hillsborough Arts Councils, Tampa Downtown Partnership, Tampa Bay Technology Forum, and the Florida High Tech Corridor.

Some three weeks later, four Tampa Bay women – Michelle Bauer, Deanne Roberts, Karen Raihill and Sigrid Tidmore – joined up with Richard Florida again at an international conference in Memphis. Over several days, the participants crafted the "Memphis Manifesto" which challenges communities to take actions to promote a creative economy based upon the three pillars of Talent, Technology and Tolerance.

By the summer of 2003, Deb Talbot joined the other four women to incorporate CreativeTampaBay with the idea that it would serve as a catalyst for perpetuating the energy and commitment of all who are eager to see Tampa Bay take full advantage of its unique heritage, geography, and economic strengths in building our own creative economy. Today, the organization reaches thousands throughout the region and is proud to have a diverse board representing creative enclaves throughout the seven counties that make up Tampa Bay.

Source: CreativeTampaBay (2007)

Table 7.1 America's most creative cities

Rank	Region	Creativity index score
1	San Francisco	1057
2	Austin	1028
3	Boston	1015
4	San Diego	1015
5	Seattle	1008
6	Raleigh–Durham	996
7	Houston	980
8	Washington	964
9	New York	962
10	Dallas–Fort Worth	960
11	Minneapolis–St. Paul	960
12	Los Angeles	942
13	Atlanta	940
14	Denver	940
15	Chicago	935

Source: Florida (2005a: 156)

Florida's advice for urban governments in need of economic revitalization begins with the question: "In a world where people are highly mobile, why do they choose some cities over others and for what reasons?" (Florida, 2005a: 33). He answers by noting that the business incentives and downtown revitalization projects used to lure companies do not necessarily attract highly creative people. Instead, such people prefer places with innovative, diverse and tolerant cultures. Therefore, Florida (2005a: 33) argues:

> While economists and social scientists have paid a lot of attention to how companies decide where to locate, they have virtually ignored how people do so. This is the fundamental question I sought to answer. In my interviews and focus groups, the same response kept coming back: People said that economic and lifestyle considerations *both* matter, and so does the mix of these two factors. In reality, people were not making the career decisions or geographic moves that the standard theories said they should: They were not slavishly following jobs to places. Instead, it appeared that highly educated individuals were drawn to places that were inclusive and diverse. Not only did my qualitative research indicate this trend, but the statistical analysis proved the same.

Though insightful, Florida's suggestion that creative people's preference for certain cities relates directly to those cities' economic growth warrants more

Plate 7.3   Not so aesthetic for creative people: Riverside Plaza in Minneapolis, Minnesota

analysis. How does a traditional city transform itself into a creative city? In his books on the creative class, Florida praises the success of a handful of creative centers, notably: Austin, Boston, San Francisco, Seattle and Washington, DC. He claims that these cities have been able to put together "the 3Ts of economic growth": technology, talent and tolerance. However, given their favorable geographical locations and local research institutions, these cities stand out from average American cities, let alone other cities around the world. Whether declining industrial towns in the American Midwest can follow suit by improving their cultural assets and art institutions is certainly debatable. The creative assets of successful cities may not be so easy to reproduce. Furthermore, based on empirical research on the impact of creative artists on urban growth, Markusen (2006) contends that the locational, cultural and political preferences of artists, a key component of Florida's creative class, differ greatly from those of scientists, engineers, managers and lawyers. As such, the task of developing an ideal creative environment for the full range of creative capital proves daunting and unlikely at best. All in all, the cultivation of a dynamic and globally successful creative city is easier said than done.

Despite these impracticalities, news magazines, including *Newsweek* and the *Atlantic Monthly*, have responded favorably to the creative capital thesis, publishing numerous short articles by or about Florida. In addition, many urban politicians circulate the creative class thesis in political speeches and apply some of its core ideas to their urban development projects. Mayors of large industrial cities in Japan, including Osaka and Yokohama, have recently launched creative city initiatives in the hope of diversifying the urban economic base and enriching the lives of their citizens through cultural activities (see Case study 7.1). The city governments of both Berlin (Kratke, 2004) and Dublin (Boyle, 2006) include the cultivation of a creative atmosphere in their development policies. Ever more city governments commission researchers or consultants to identify the advantages of attracting creative capital, as well as their cities' challenges to becoming creative centers. One such example includes Toronto's "Strategies for Creative Spaces and Cities Project," conducted by the Program on Globalization and Regional Innovation Systems at the University of Toronto. In its creative city endeavor, according to the program, Toronto should focus on 14 strategic missions, including increasing funding for arts and cultural institutions (Table 7.2).

Case study 7.1
Creative cities in Japan

The highly urbanized country of Japan boasts one of the most developed economies in the world. However, apart from Tokyo, few Japanese cities have produced a substantial amount of academic literature (in English, at least) on their economic restructuring, development policies or urban decline. Considering the enormous scholarly attention devoted to the Japanese economy and to Japanese companies, the lack of empirical and theoretical efforts better to understand urban economy in Japan is striking. It is not uncommon to see certain cities "over-researched" at certain times by urban scholars. For example, Chicago was fairly overexploited by American urban sociologists in the first half of the twentieth century. Manchester represents a more recent example of a city endlessly dissected to produce a large body of academic journal articles, books and reviews on its urban entrepreneurialism. But for every over-researched city, there are many more under-researched cities. Most large industrial cities in Japan have not received their due attention.

As late-comers in industrialization, Japanese cities in general have not suffered the same aftermath of deindustrialization as British cities in the Midlands and American cities in the Northeast. However, this does not mean that they have been completely exempted from the urban impact of industrial relocation, outsourcing and unemployment. Instead, assorted problems go overlooked in Japanese urban economy (Sasaki, 2004), revealing the Anglo-American bias in urban studies.

Yokohama and Osaka are the second- and third-largest cities in Japan, respectively. Both are seaports and, in addition to trade, their economies have been established around heavy industry. However, they have both witnessed the gradual weakening of their local industries. In order to attract creative talents and diversify their economic bases, both cities have recently launched a creative city initiative. "The Creative City Yokohama" was set up in 2004 to develop a cluster of creative industries with a specialty in visual culture (Figure 7.1). Yokohama has a government office, named the

Creative City • Yokohama

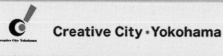

The city of Yokohama has taken the realizati of this "Creative City" vision as a key pillar of our strategy for the city's development.

To pursue an autonomous advancement whi enriching the civic life, the vision will instill Yokohama with new value and appeal. To this end, it will make full use of the resources accumulated during our unique history as a city that grew up around its port, and actively exercise our creative potential in the spheres of art and culture.

MOVIE

The realization project of the "Creative City Yokohama".

National Art Park Plan	Formation of Creative Core Areas	Image Culture City	International Triennale of Contemporary Art Yokohama	Nurturing of Future Creators

| Grand design for central waterfront area to build a creative city. | Diversion of local resources to create a space for the creation, exhibition, revitalize central neighborhoods. | Formation of Asian video hub through clustering of video content producers, etc. | Staging of international contemporary art exhibitions to raise Creative City Yokohama's profile. | Develop a wide range of human resources to support culture and art. |

Figure 7.1 Creative City, Yokohama

Source: **Official website of the City of Yokohama**

Creative City Headquarters, designated to revitalize the city through promoting cultural and artistic activities (official website of the City of Yokohama, 2006).

The Mayor of Osaka has recently announced his creative city initiative, in an attempt to regain the support of his constituency after a series of scandals in 2005. The city hosted a large international symposium on "Building a Creative City." Based on the outcomes of this symposium, the city government is currently working on creative urban spaces as well as creative human resources (official website of the City of Osaka, 2006). The Osaka Urban Revitalization Task Force has been created to co-ordinate and promote various projects aimed at reinventing Osaka as a "City of Creativity." In addition, Osaka City University has created a new graduate school program specializing in the creative city. As with most of its economic development measures, the Japanese government, whether at the national, provincial or city level, seems to be fully in charge of the creative city initiative.

Box 7.3 Osaka's creative city initiative

Although reshaping the city management system has been the focus of administrative reforms, we feel we must now start working on developing new strategies to form a vibrant and interactive "Creative City," or a city able to produce new, innovative industries and culture by drawing on individual potentials. Implementation of such strategies is to begin in the next fiscal year. Creative human resources is the key to a Creative City, and in order to draw such a pool of talent, we will formulate a strategy based on an assessment of issues at hand and researching ways to resolve them. For this purpose, a "dream team" of young city employees will be established. They will benefit from the expertise of Creative City researchers and also incorporate various voices from citizens and NPOs [non-profit organizations]. City personnel can directly e-mail me their ideas and suggestions on this theme. In addition, our new venture system will also allow personnel to follow through and act upon suggestions they have made. During the first half of this fiscal year, we will put together these ideas and compile a Creative City Strategy by the end of March 2007 for real implementation starting in FY 2008.

(Mayor's press conference, April 26, 2006)
Source: Official website of the City of Osaka

Table 7.2 Toronto's recommended strategies for a creative city

Project categories	Strategic opportunities for action
People	Ensure that all youth are encouraged to think creatively. 1 Expand creative programming for youth 2 Transform local community 3 Fund arts and creativity in public education
Enterprise	Create the conditions that allow enterprise and their financiers to take risks; Increase support for sectors that are gaining international attention; Inspire all firms and entrepreneurs across all sectors to think creatively. 4 Provide specialized entrepreneurship support/business skill development for creative industries 5 Increase available cultural/creative "risk" capital 6 Advance Toronto as a Centre of Design 7 Develop creativity/innovation convergence centre
Space	Achieve a balance between the need for major iconic cultural institutions and supporting grassroots; Design space that is affordable and sustainable for creative work and play. 8 Provide affordable and stable creative space systematically 9 Create a mortgage investment fund for creative development 10 Support development of waterfront ground-floor strategy 11 Support design review panel 12 Animate the City Below – Toronto Ravines
Connectivity	Provide organizing infrastructure that will connect existing creative activity and resources that currently work in silos. 13 Develop new infrastructure dedicated to connecting and promoting creative Toronto 14 Provide ongoing, stable funding for creative projects

Source: Program on Globalization and Regional Innovation Systems, University of Toronto (2006)

Criticisms of Florida's creative class thesis come from many different directions, as he notes in his latest book (2005a). The lack of academic rigor and sophistication has been obvious in the thesis (Markusen, 2006). Jamie Peck (2005: 740) sums up many academics' assessments:

> The thesis – that urban fortunes increasingly turn on the capacity to attract, retain and even pamper a mobile and finicky class of "creatives," whose aggregate efforts have become the primary drivers of economic development

– has proved to be a hugely seductive one for civic leaders around the world, competition amongst whom has subsequently worked to inflate Florida's speaking fees well into the five-figure range.

As was shown in Chapter 6 on urban economic development, the creative city idea seems to be the latest policy phenomenon that reflects neo-liberal, business-friendly, upper-middle-class-oriented views of urban governance. Peck (2005: 740–1) goes on to say:

> In the field of urban policy, which has hardly been cluttered with new and innovative ideas lately, creativity strategies have quickly become the policies of choice, since they license both a discursively distinctive and an ostensibly deliverable development agenda. No less significantly, though, they also work quietly with the grain of extant "neoliberal" development agendas, framed around interurban competition, gentrification, middle-class consumption and place-marketing – quietly, in the sense that the banal nature of urban creativity strategies in practice is drowned out by the hyperbolic and overstated character of Florida's sales pitch, in which the arrival of the Creative Age takes the form of an unstoppable social revolution.

So why has the notion of the creative class become a policy phenomenon for city governments? Richard Florida's skills of self-promotion have helped. The surrounding media frenzy fuels political interest in the image of a cool, creative citizen base. Politicians, as usual, capitalize on optimistic views of the urban economy, no matter how impractical those views are. Or, as is suggested in "new realism" (Quilley, 2000), the dire economic situation in declining industrial towns might force their mayors to focus less on public schools or housing and more on attracting the creative people who might drive up the value of the local property market.

The restless and transnational nature of urban economic change requires city officials to seek a new vision of the urban economy. This parallels the way corporate executives try to redefine the larger industries in which their companies belong, often exaggerating the volatility and globality of the current market situation in order to advance priorities in their corporate management.

Conventional wisdom states that the vibrancy of a downtown business district reflects the overall economic conditions of the city. In contrast, a new thinking states that an urban atmosphere offering iconic architecture, cultural amenities and diverse lifestyles represents a prerequisite for urban economic development (Gospodini, 2002; Sklair, 2005). This new interpretation of the relationship between urban economy and urban culture and design helps to explain why so many medium-sized cities have courted big-name architects and hosted events to celebrate multicultural lifestyles.

In today's competitive world, city governments are expected to be aggressive and, more importantly, creative in their efforts to achieve economic development. A well-designed package of business incentives may boost one city's chances of attracting private capital investment, but when most other cities offer the same, such initiatives may not stand out. Ever more urban politicians talk about the economic value of culture, the arts, history and green space, as well as that of high-tech industries. Such politicians focus on these traditionally non-economic factors not only because of their direct contribution to urban economic development but because of their role in attracting talents who might play a key role in future economic growth.

Further reading

Florida, Richard, 2002, *The Rise of the Creative Class: And How It's Transforming Work, Leisure, Community and Everyday Life*, New York: Basic Books. A seminal book that has been at the center of the debate over the creative class thesis. Florida claims that the contemporary world is witnessing the emergence of a new kind of economy that is led by a new social class – the creative class.

Peck, Jamie, 2005, "Struggling with the Creative Class," *International Journal of Urban Regional Research*, 29(4), pp. 740–770. Peck criticizes the creative city idea as it reflects neo-liberal, business-friendly and upper-middle-class-oriented views of urban governance.

Scott, Allen J., 2000, *The Cultural Economy of Cities: Essays on the Geography of Image-Producing Industries*, London: Sage. This collection of Scott's empirical studies examines the clustering of "cultural-products sectors" and its contribution to the urban economy, with special attention to Los Angeles and Paris as the two primary global centers of cultural industries.

Throsby, David, 2001, *Economics and Culture*, Cambridge: Cambridge University Press. This book provides an excellent review of the cultural economics literature, which explores the cultural context of economics as well as the economic dimensions of culture.

Zukin, Sharon, 1995, *The Cultures of Cities*, Cambridge: Blackwell. In this collection of essays, Zukin looks in detail at the use of cultural strategies in the urban redevelopment of America's declining or economically undervalued communities.

Part Three

The national economy and capital cities in developing countries

Large cities in developing countries have been understood and examined through the notion of "third world urbanization." Generally, the third world urbanization literature views such cities as problematic and plagued with a series of crises, including: economic hardship, government failure, historical conflict and natural disaster in the urban South (Robinson, 2006; Simone, 2004). David A. Smith (1996: 2) notes that "the sights, smells, sounds and faces" of Calcutta, Jakarta, Lagos, Lima and other rapidly growing cities of poor countries have dominated academic writings on urban changes in Africa, Asia and Latin America. Alongside these sensory-rich empirical accounts of urban problems, Marxist urban critiques have pervaded studies of third world cities since the early 1980s. Influenced heavily by the dependency/world-system perspective, Marxists, through comparative analyses of third world cities, have pointed to dependent or peripheral urbanization as a major cause of massive poverty (Roberts, 1978; Timberlake, 1985).

Pessimism aside, such third world urbanization literature has suffered a fundamental disconnection from some of the latest theoretical developments in urban studies. Although third world urbanists have led the field in viewing urban changes from a global perspective (Davis and Tajbakhsh, 2005; Smith, 2003), their efforts have not been translated into more recent researches on the urban impact of globalization. Simply put, their work has not received due credit by the newly emerging world-cities research, in which third world cities remain largely invisible. Part Three deals with urban economic change in the developing world and related academic views and writings.

In Part Two, we examined various urban policies attempted by city governments in the developed world with the goal of overcoming the effects of deindustrialization and eventually improving their cities' competitiveness in a globalizing economy. Governments of large cities in the developing world also attempt to overcome various internal and external constraints, but different ones from those in the developed world. It is difficult to gauge which of the two types of government face the more difficult task. Urban scholars have so far struggled to produce comparative research on urban policies across the Atlantic (Savitch and Kantor, 2002), making it almost impossible to undertake a comparative analysis of large cities in the developed world and those in the developing world.

Chapter 8 reviews existing studies of large cities in the developing world and their urban problems. Chapter 9 examines the world-city projects that many national governments of the developing world have launched in the hope of "putting their cities on the map." In Chapter 10, we critically investigate the uneven nature of globalization through growing inequalities in the large, globalizing cities of the developing world.

8 Third world cities

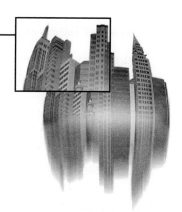

Learning objectives

- To have a critical understanding of the studies of third world cities
- To understand population issues and policies in mega cities of the developing world
- To explore the positive and negative effects of informal economies in third world cities
- To describe major characteristics of primate cities

The common use of the term "Third World" reflects a lingering effect of cold war politics on underdeveloped countries. As the cold war ended more than 15 years ago, the term has become irrelevant and largely replaced by less politically charged terms, such as the developing or less developed world, emerging markets and the Global South. However, when addressing large cities in developing countries, the term "third world urbanization" comes in handy. It is still used widely in academic writings to conceptualize various urban issues in those mega cities, ranging from squatters and overpopulation to street vendors and the informal sector to polluted rivers and environmental sustainability.

In this chapter we review the existing studies of third world cities to assess what has been written about their nature and their positions, in both the world and their national economies. There is no single defining factor of third world city-ness. A combination of assorted urban problems observed in most large cities of poor countries has been used to define third world cities, much as the shared economic characteristics of London, New York, and Tokyo have been used to define world city-ness in world-cities research. This chapter examines the traditional notion of third world cities, specifically looking at the conceptualization of mega cities and primate cities, while the following two chapters will highlight recent urban changes observed therein. The first section of Chapter 8 reviews studies of third world cities, which pre-date the recent explosion of writings on world cities and their role in a global economy. In the second section, we identify the most pronounced urban problems of the developing world, in which a disproportionate share of the world's mega cities are located. The third and final section looks at primate cities and their role in the economic growth of their nations.

Third world cities in urban studies

Third world cities have been grossly under-theorized and frequently under-emphasized in mainstream urban studies. As a result, many urban scholars – and third world urbanists in particular – have called for serious scholarship on them (Robinson, 2006; Smith, 2003). While the vast majority of urban dwellers are found in third world cities, their urban experiences remain almost invisible in key theories or concepts in urban studies, which instead show a clear bias toward the cities of developed countries. If geographical imbalances continue, both in the production sites of urban theories and in the populations of large cities, the current field of urban studies may simply lose relevance.

Why does the "under-theorized and under-emphasized" problem persist, even though it has been widely recognized? Despite criticism of the so-called lack of research on non-Western cities, there already exists a sizable body of literature on large cities in the developing world. The more likely problem lies in the fact that much of the literature on the developing world exhibits a strong empirical orientation and, therefore, a repetitive tendency in terms of research questions and method. Simply put, the field needs more concerted efforts to theorize urban economic changes in third world cities.

Another major problem in the studies of third world cities is a tendency among urban scholars to de-emphasize or even ignore myriad personalities and characteristics that individual third world cities have developed, focusing instead and almost exclusively on the problematic aspects of those cities. Third world cities continue to be analyzed through their relative lack of modernity, industrialization, public services and, more recently, urban sustainability. Few academic efforts highlight or theorize on cosmopolitan cultures, vibrant communities or urban politics in the global urban South, perpetuating the assumption that such issues should be reserved for the more developed North.

The weak presence of the third world urbanization literature in contemporary urban studies should also be attributed to the highly limited geography of recent world-cities research. Since this has focused almost exclusively on the "command

and control centers" of a globalizing world economy, many third world cities, if mentioned at all, are portrayed as places significantly marginalized from the process of economic globalization (Grant and Short, 2002; Gugler, 2004; Robinson, 2002). Their centuries of involvement with the wider world and the resultant changes in their urban economies have been simply ignored in most recent studies of world cities and globalization.

It is interesting to see how third world cities have been viewed and examined by so-called "first world" urban scholars over the years. The fact that the vast majority of empirical work in urban studies relates to either the United Kingdom or the United States should not negate the enormous achievements that studies of "Other" cities have made in expanding and deepening our understanding of the relationship between cities and economies. Though always in the minority, many urban scholars in earlier years wrote and taught about the urban histories, economic bases and urban policies of various under-represented developing cities that provide a better grasp of the multiplicity and complexity of urbanization processes. Unfortunately, this body of research remains small, and its findings contribute little to the theorizing work of urban changes.

A large number of studies on third world cities have been conducted and published in various linguistic and national contexts, many of which prove more informative than those conducted by non-local researchers. However, we limit our review to studies of large cities in the developing world that have been documented in English.

The questions that we attempt to answer include:

- How and in what context did American and European urban scholars realize the importance of examining cities in the underdeveloped world of Africa, Asia and Latin America?
- How have their assumptions and research questions changed in their investigation of urban change in third world cities?
- What are the latest trends in the study of third world cities?

According to Diane Davis's (2005) review of the history of American urban sociology, US sociologists did not begin to study cities outside their national borders until the mid-1950s. Many of the leading American sociologists of the time were instead occupied with urban problems and industrial growth in domestic cities. In 1955, the *American Journal of Sociology* devoted a special issue to "world urbanism," emphasizing the need for more intensive research into non-Western cities and into comparative international urbanism in general (Hauser, 1955). The issue includes three empirical studies of specific areas: great cities in Southeast Asia (Ginsburg, 1955), population growth in Indian cities

(Crane, 1955) and traditional farming communities in Yoruba cities of Nigeria (Bascom, 1955).

These early writings offer a glimpse of both the limitations and the potential of the third world-cities research of the 1950s. All three authors seem exceptionally knowledgeable about their respective case study cities, setting a good example for future generations in the discipline. Furthermore, their enthusiasm for researching non-Western cities cannot be mistaken. However, the similarity stops there. Ginsburg (1955) highlights two main characteristics of large cities in Southeast Asia: primate cities and colonial cities. Comparing those cities to urban centers in the US, he notes that, "to a very large degree, in Southeast Asia growth is primarily in the great cities, where in the United States in the last three decades at least, growth has been more rapid in the small and intermediate cities" (p. 457). As for the colonial history of Southeast Asia, he states that, "more important, the great cities were, and to a large degree still are, foreign creations, essentially alien to the Southeast Asian landscape. Although several began as native settlements, they developed as foreign exclaves on the margins of Asia, in much the same way as did the plantation system and mining operations" (p. 458).

In his account of urban development in India, Crane (1955) places great emphasis on population problems in large Indian cities. He points to administration, handicrafts and trade as major causes for development in pre-colonial Indian cities, but "since the Europeans came, it has been modern industry" that has accelerated urban growth (p. 463).

Compared to these two, Bascom (1955) seems more eager to revise the existing notion of urbanism that is founded in Western urban experiences. In the Yoruba of west Nigeria, large, dense and permanent settlements are economically based on farming rather than manufacturing industries. He argues that these settlements are in fact urban, despite their lack of industrial bases or the social heterogeneity defined by Louis Wirth as a critical element of urbanism. Bascom concludes his fascinating exploration of Yoruban cities (p. 453) with its implications for future studies:

> It is necessary at least to distinguish between industrial and nonindustrial cities and between cosmopolitan and noncosmopolitan cities. It is also suggested that the existence of a formalized government which exercises authority over the primary groups and incorporates them into a political community may be more useful than heterogeneity when applied cross-culturally, since it is less subjective. It is hoped that these points may broaden the concept of urbanization so that it is less dependent upon the historical conditions of western urbanization and so that it can be applied more profitably to the study of the urban centers of India and Southeast Asia.

While Ginsburg recognizes characteristics of American frontier towns in Southeast Asian cities, Crane explores causes of over-population in Indian cities. In the meantime, Bascom views Yoruban settlements as neither an earlier version of Western cities nor as some uncontrollable problem, but as a new type of urbanization that warrants closer examination. These not-so-subtle differences in how Western urban scholars approach the developing world and their major cities persisted over the next five decades, although the cultural and political biases that earlier studies reveal have abated significantly.

Although such broad categorization runs risks of over-simplification and omission, for the sake of comparison we can divide the third world urbanization literature of the past fifty years into four general groups: empirical accounts of urban problems; urban spatial analyses of third world cities; international political economy approaches to underdevelopment; and finally recent literature on the effects of globalization and neo-liberalism on the social and economic fabric of large cities in the developing world. We review each in turn.

The first and the most established tradition in the study of third world cities is to examine urban problems and attempts to solve them. This tradition has shown a strong empirical orientation. The rich description of case study areas, based on researchers' extensive time and efforts in the field, emerges as the undeniable strength of this category of literature. Some focus on one or more cities within a country or region to review their historical development, including pre- and post-colonial periods (Amirahmadi and El-Shakhs, 1993; de Blij, 1968; Harris, 1978; Tarver, 1994). Meanwhile, others cover a wide range of cities with diverse urban experiences (Bryceson and Potts, 2006; Gugler, 1988 and 1996; Kasarda and Parnell, 1993; Oberai, 1993; Rakodi, 1997; Seabrook, 1996). For example, one edited book on African cities (Rakodi, 1997) addresses challenges of urban Africa, ranging from overpopulation to residential property markets to health problems. Such edited books, which offer extensive coverage of cities and their problems, have helped to shape a widespread perception in urban studies that third world urbanization literature excels in empirical detail, but lacks theory.

The second group of studies represents a rather short-lived tradition of the 1960s and 1970s when some academics (mostly geographers) attempted to draw generalized spatial patterns of urban structure in the underdeveloped world (Friedmann, 1966; McGee, 1967; Taaffe et al., 1963). Not all of them used the then newly emerging spatial modeling methods, but some did, in order to test their sophisticated analysis techniques in third world cities. Pointing out the fact that almost all existing urban models, such as the concentric model and the central place theory, were based on Western urban experiences, they attempted to generalize distinct land-use patterns and residential segregations across the third world (Lowder, 1986). Taaffe *et al.*'s work (1963) on Ghana and Nigeria helped to illuminate the impact of colonial transportation networks on national urban hierarchies, as well as the roles of major African cities in the international context.

The third tradition refers to an academic endeavor in the 1980s to interpret third world urbanization in the context of the capitalist world system, led primarily by students of dependency theory and world system theory (Chase-Dunn, 1985; Potter, 1992; Roberts, 1978; David A. Smith, 1996; Timberlake, 1985). Placing major cities of developing countries in the international division of labor, Marxist scholars link problems of poverty, the informal sector and overpopulation to the imposed structural features of an inherently unequal capitalist world economy. In other words, they reject the idea that various urban problems observed throughout the South attest to a socio-political failure, by governments or their populaces, to embrace Western technologies, values and economic systems, as suggested by modernization theorists. Instead, they posit that underdevelopment in third world cities has resulted from the unfairly structured trading systems and production networks between the developing and the developed worlds. Building on Andre Gunder Frank's "satellite–metropole relations" in the world economy (Figure 8.1), terms like "dependent urbanization" and "(semi-)peripheral urbanization" have been used to theorize urban problems and transitions in the poorest and most populous cities of the world. While being praised for taking an international political-economic approach to urban poverty issues in third world cities, dependency and world system theorists have been criticized for their neglect of diversity within the developing world. As they put too much emphasis on convergences between third world cities, they have almost lost sight of divergences originated in place-specific characteristics.

The fourth and the most recent tradition of third world cities research examines the impact of globalization on major cities of the developing world (Grant and Nijman, 2002; Grant and Short, 2002; Gugler, 2004; Keyder, 2005; Logan, 2002; Roberts, 2005; Segbers, 2007; Ward, 1998). Built on decades of accumulated knowledge of third world cities, collected from diverse academic traditions, this latest trend engages in various theoretical debates on urban studies and the social sciences in general, and offers strong empirical groundings. Working to fill in the theoretical and empirical gaps between third world-cities research and world-cities research, they argue that the geography of world-cities research must be more globalized to include Africa, Asia, Latin America and other less researched parts of the world. In their comparative historical analysis of Accra and Mumbai, for example, Grant and Nijman (2002) identify the emergence of different business centers responding to different global forces, such as past colonialism and more recent foreign investment. Keyder (2005) focuses on the growing trend of social polarization in Istanbul as the city undergoes a massive spatial restructuring under

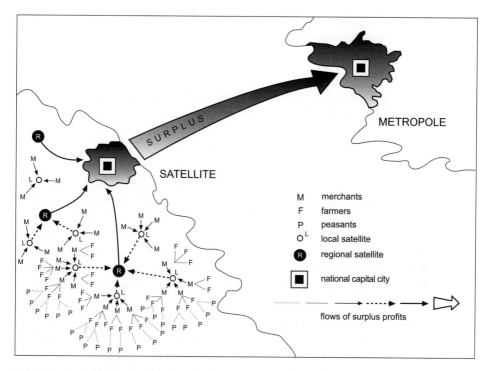

Figure 8.1 Satellite–metropole relations in the world economy

Source: **Potter** *et al.* **(1999: 65)**

neo-liberal, globalization-friendly economic policies. Another example of the new studies of third world cities can be found in Jennifer Robinson's (2006) work on Johannesburg, a city laden with overwhelmingly negative stories. Taking on both globalization and the development of its ragged economy, notes Robinson, Johannesburg works to build an African world-class city which preserves its African cultural integrity and utilizes its network of African connections.

Although still prominent, the gap between third world-cities research and world-cities research has narrowed as the body of literature examining the interplay of globalization and non-Western cities expands. A growing number of edited books focus on the urban impact of globalization in large cities across the developing world (Grant and Short, 2002; Gugler, 2004; Öncü and Weyland, 1997). In addition, some publishers have launched new series on the world cities of Buenos Aires (Keeling, 1996), Havana (Segre *et al.*, 1997) and Tehran (Madanipour, 1998), whose documented historical relations with the outside world were not readily available in English before.

David A. Smith (2003: 122), whose take on "urbanization in the world economy" perspective certainly pre-dates the current world city system research, offers highly valuable insights into the renewed connection between world-cities research and third world-cities research:

> While the notion of a "global city system" may imply a return to a focus on those "chain(s) of metropoles and satellites" reaching deep into the peripheral regions of the world-economy that A.G. Frank described many years ago, in fact the swelling cities of the poor in "the South" tend to be both under-theorized and understudied. For a variety of reasons, the focus of much of the "global cities" research is on the dominant command centers near the apex of the urban hierarchy; attention falls off as we move toward cities near the bottom. The unfortunate result is that the massive changes and growing problems of these megacities are only addressed in descriptive studies based on theoretically eclectic frameworks (Gugler 1996, 1997). The careful attention on the urbanization/underdevelopment dynamic, stressed by the urbanization in the world-economy's notion of "dependent urbanization," is missing. The challenge to revitalize this neglected intellectual space is particular pressing.

Mega cities in the developing world

The past few decades have witnessed a sharp contrast in population trends between major cities in the developed world and those in the developing world. While populations decline in developed big cities through decentralization and suburbanization, large cities of developing nations have seen their populations grow through centralization and rural–urban migration – although many non-economic causes, such as political conflicts and natural disasters, also contribute to the rapid population growth in some third world cities. As a result, most of the largest and fastest-growing cities can be found in the developing world. What is more, their populations grow in direct proportion with the wholesale expansion of urban slums, one of the most threatening features of the global urban South (UN-HABITAT, 2003). This section looks at the growth of mega cities, truly large cities with more than 10 million inhabitants, in the developing world.

Table 8.1 lists the cities that have already reached, or are projected to reach within the next decade, the staggering population figure of 10 million. It is more accurate to describe these mega cities as city regions since the expansion presses outwards as well as upwards. There are some urban areas that act as coherent city regions but fail to claim this distinction because urban growth crosses different administrative boundaries (Hall and Pain, 2006). For example, at the beginning of the twenty-first century, the administrative area of Greater London has a population of approximately 7.5 million, but the population of the wider metropolitan areas reaches between 12 and 14 million. Table 8.1 tends to ignore city regions like

Table 8.1 The world's mega cities with 10 million people or more

Rank	1950 City	Population	1975 City	Population	2005 City	Population	2015 City	Population
1	New York–Newark	12.3	Tokyo	26.6	Tokyo	35.3	Tokyo	36.2
2	Tokyo	11.3	New York–Newark	15.9	Mexico City	19.2	Mumbai	22.6
3			Shanghai	11.4	New York–Newark	18.5	Delhi	20.9
4			Mexico City	10.7	Mumbai	18.3	Mexico City	20.6
5					São Paulo	18.3	São Paulo	20.0
6					Delhi	15.3	New York	19.7
7					Kolkata	14.3	Dhaka	17.9
8					Buenos Aires	13.3	Jakarta	17.5
9					Jakarta	13.2	Lagos	17.0
10					Shanghai	12.7	Kolkata	16.8
11					Dhaka	12.6	Karachi	16.2
12					Los Angeles	12.1	Buenos Aires	14.6
13					Karachi	11.8	Cairo	13.1
14					Rio de Janeiro	11.5	Los Angeles	12.9
15					Osaka–Kobe	11.3	Shanghai	12.7
16					Cairo	11.1	Metro Manila	12.6
17					Lagos	11.1	Rio de Janeiro	12.4
18					Beijing	10.8	Osaka–Kobe	11.4
19					Metro Manila	10.7	Istanbul	11.3
20					Moscow	10.7	Beijing	11.1
21							Moscow	10.9
22							Paris	10.0

Source: United Nations (2005)

London. But even with this understanding, the developing world clearly dominates the recent growth of very large city regions. In 1950 only the populations of New York and Tokyo had passed the 10-million mark. By 2005, the list of mega cities had increased to 20, 16 of which were located in the developing world (Figure 8.2).

Furthermore, the UN-HABITAT (2006–7) sees a new trend in urbanization, as "meta cities," massive conurbations of more than 20 million people – above and beyond the scale of mega cities – are now gaining ground in Africa, Asia and Latin America. The list of potential meta cities by 2015 includes Mumbai, Delhi, Mexico and São Paulo as well as Tokyo, the world's most populous city for the past three decades. The dynamics of population growth in meta cities and mega cities result largely from high levels of rural–urban migration. Large urban agglomerations in the developing world serve as magnets for large-scale movement from the country to the city.

Savitch and Kantor's (2002) fourfold typology of the relationships between population and economic change helps us understand population-related problems in mega cities in the developing world (Table 8.2). In the developed world, many former industrial cities experience economic distress and population loss, while some booming high-tech cities, such as Austin, Texas, simultaneously enjoy both fast population growth and economic prosperity. Mega cities in the developing world, in contrast, experience large population gains without the rapid economic growth needed to support the increased population. This is labeled "impaction" in the table, as it occurs when people move off rural land in search of better opportunities and fail to find them. Migration into the largest city in the country is often accompanied by poorer living conditions and unemployment, resulting in "impoverished growth." Most mega cities in the developing world have experienced impoverished growth for decades, prompting their national governments to implement assorted "national urban development strategies," aimed solely at curbing rural migration flows to leading cities (Richardson, 1981).

The sheer size and growth rates of mega cities tend to overwhelm the carrying capacity of urban infrastructure. The inability of housing markets to meet the ever-growing demand leads to the massive expansion of slum settlements (Plate 8.1). Slums emerge as a distinct and dominant form of settlement in mega cities of the developing world. Table 8.3 shows that more than 2.9 billion people currently reside in urban areas, almost 2 billion of whom are in cities of the developing world. Of those urban residents in the developing world, almost 850 million, 42.7 percent, are indeed slum-dwellers. UN-HABITAT (2006–7) predicts that the world slum population will reach the 1-billion mark in 2007 and that the vast majority will be residents of third world cities. Urbanization has become virtually

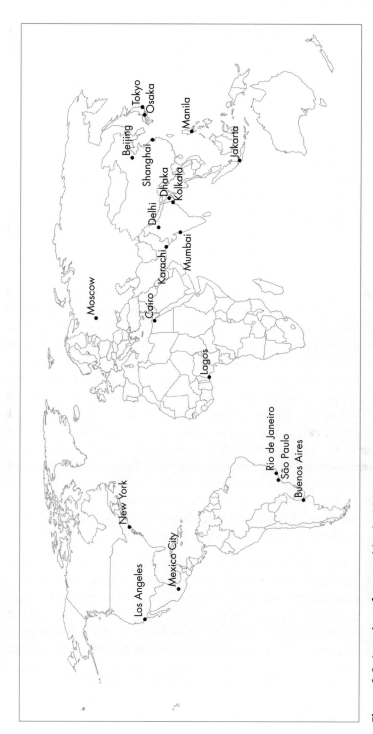

Figure 8.2 Location of mega cities in 2005

Source: **United Nations (2005)**

Table 8.2 A typology of economic and population changes in cities

	Economic change	
	Prosperity	*Distress*
Population		
gain	growth (new-age boom towns, high-tech corridors, edge cities)	impaction (impoverished growth, squatter villages)
loss	dedensification (renewed CBD, luxury high-rises, gentrified and/or stable neighborhoods)	decline (hollowed-out CBD, derelict, high-crime neighborhoods)

Source: Savitch and Kantor (2002: 10)

Plate 8.1   Deteriorating housing conditions in Kolkata (Calcutta), India

Table 8.3 Urban and slum population in the world, 2001

	Total population (thousands)	Urban population (thousands)	Percentage urban	Slum population (thousands)	Percentage slum
World	6,134,124	2,923,184	47.7	912,918	31.2
Developed Regions	985,592	753,909	76.5	45,191	6.0
Developing Regions	4,865,893	1,988,093	40.9	849,013	42.7
CIS*	282,639	181,182	64.1	18,714	10.3

Note: * The Commonwealth of Independent States, consisting of 11 former Soviet republics

Source: UN-HABITAT (2006–7)

synonymous with slum growth in the poorest parts of the world, especially in sub-Saharan Africa and South Asia, where annual slum and urban growth rates are almost identical.

Along with slum settlements, the expansion of informal sectors represents another defining feature of a mega city with a large and fast-growing population. The informal economy can be roughly defined as a group of economic activities not regulated by the government. Such activities range from illegal, criminal jobs (narcotic traffickers or prostitutes) to small-scale manufacturers and retailers (such as street vendors) to garbage collectors, home builders and tool repairers – yet, most academics distinguish the unregulated production and distribution of otherwise licit goods and services from those conventionally labeled criminal activities (Table 8.4). The informal sector is not autonomous from the formal sector, as the two sectors intersect through subcontracting, franchising and labor and capital flows (Potter and Lloyd-Evans, 1998). Indeed, the boundaries of the informal economy substantially vary in different societies and historical circumstances, as informality reflects not so much the intrinsic characteristics of activities as the social definition of state intervention (Castells and Portes, 1989). Even though informal sector employment exists in any and all types of society, it especially flourishes within societies marked by underdevelopment and, more specifically: poor governance, lagging modern private businesses, high unemployment rates and scant or non-existent social safety nets in African, Asian, and Latin American societies.

Table 8.4 Types of economic activity

Process of production and distribution	Final product	Economy type
licit	licit	formal
illicit	licit	informal
licit or illicit	illicit	criminal

Source: Castells and Portes (1989: 14)

In fact, the informal sector accounts for about a quarter of urban jobs in Latin America, 61 percent in Sub-Saharan Africa and 40 to 60 percent in Asia (excluding relatively developed East Asian countries) (Sethuraman, 1997). Some Asian cities report strikingly high estimates of informal employment: 65 percent in Dhaka, 65 percent in Jakarta, 50 percent in Metro Manila and 49 percent in Bangkok. What is worse, these numbers have been growing quickly in recent years, thanks to poor governance and the resultant fragility of the public and formal sectors.

There has long been a divide among policy-makers and academics regarding how governments should approach the informal sector. Some argue such numbers as those presented above offer overwhelming evidence of a correlation between informal employment and urban poverty. In other words, governments of third world cities should tackle the informal sector, since it undermines the capacity of the formal sector, namely the state and the modern private sector, to generate adequate employment and tax revenues. Others make a counter-argument that, since small-scale industrial, handicraft and repair establishments make up the core of the informal sector, governments should support these enterprises with technical assistance, training and credit. The International Labor Organization (ILO) has been an ardent advocate of sustained and increased governmental support for informal work, as it sees informal occupations as ways for the poor to get by when neither corporations nor governments can provide sufficient employment for the expanding population (Bromley, 1993). De Soto (1989) points to the high productivity of the informal sector and its positive role in generating incomes and satisfying the basic needs of the poor, although his policy suggestions of deregulation and privatization are quite the opposite of the ILO's.

In reality, most governments of developing countries have attempted both to police and to assist informal economies. They do not want to lose control over the economies of their leading cities, which have become increasingly informalized. At the same time, they see the informal sector creating jobs for poor residents who otherwise would be unemployed. With a little help from the government, petty enterprises in slum areas could create considerably more jobs for slum-dwellers and especially newly arrived rural migrants who might not have many marketable skills.

Of late, however, neo-liberal policies have imposed significant reductions in state capacity both in regulating the informal sector and in subsidizing it. This trend, accompanied by massive cuts in public expenditure, has pervaded most developing countries, particularly the poorest of the poor in Africa. As the public sector continues to falter in sub-Saharan Africa, according to King (2001), many formal workers, such as public schoolteachers, need to find second or even third jobs in the informal sector to earn living wages. This may not sound so bad, as some formal sector workers take on additional income-generating activities in difficult times. However, as ever more schoolteachers take on jobs outside the schools, the quality of teaching in those schools might suffer. If that is the case, the expansion of the informal sector in mega cities may hurt, not help, the formal sector and the future of the society in general.

Informal economies and shanty towns rise in response to market failures and weakly funded public programs. The rate of population growth, in association with lax environmental controls, also generates mounting concerns over environmental pollution and public health (Plate 8.2). Water and air pollution most seriously afflict the poorest neighborhoods of rapidly growing mega cities. The lack of clean water, for example, is the main cause of infant mortality in the mega cities of the developing world.

Plate 8.2
Water pollution
in Delhi, India

Case study 8.1

Mega cities: decoding the chaos

The mega cities of the developing world are new, strange things. Their sizes and densities stagger the mind. Mumbai, for example, now holds 18 million people, two-thirds of whom live in neighborhoods where population density reaches a million people per square mile (Mehta, 2004). Lagos is another prime example of a mega city in the developing world. In 1950, when it was the capital of the British colony of Nigeria, the population was less than 300,000. By 2010, the population is estimated to reach 20 million, making it one of the largest cities in the world. Migration from throughout West Africa keeps the city's average annual growth rate at 6 percent. Over 500,000 people move to Lagos every year, a wave of rural–urban migration that seems unlikely to subside.

This level of immigration outweighs the ability of a formal economy or housing market to accommodate the city's population. The result is a city dominated by the informal and illegal economy and squatter settlements. While some see the city as a stunning example of ordinary people's ability to cope and as a source of endless entrepreneurial activity (Neuwirth, 2004), others see it differently: "The vibrancy of the squatters in Lagos is the furious activity of people who live in a globalized economy and have no safety net and virtually no hope of moving upward" (Parker, 2006: 65).

The rate of growth has, in a way, overwhelmed the city. Unlike many cities where slums are distinct from elite areas, illegal settlements and informal markets have engulfed most of this city. The problems are exacerbated by its political system. A series of kleptocratic regimes has impoverished the entire nation. Wealth is unevenly distributed while the incomes of those in the middle shrink as costs outpace wages. Corruption is rife and civil society is dominated by economic relations where the powerful and rich control the weak and poor. Most people work for someone else, paying him or her a cut from his or her income.

Conflicting interpretations abound. Lagos as a new urban order: people making their way without the benefit of a functioning urban government. Lagos as dystopia: massive population growth, paired with a shrinking economy and an inefficient public sector, lead to urban destitution. As one local politician noted, "[W]e're sitting on a powder keg here. If we don't address this question of economic growth, and I mean vigorously, there is no doubt as what's going to happen here eventually. It's just going to boil over" (quoted in Parker, 2006: 75).

It naturally follows to ask which policy measures have been proposed to tackle the issue of population growth in third world cities. International development agencies, such as UN-HABITAT, have long asserted that population growth in third world cities is excessive, and that it has been caused, or at least aggravated, by the failure of governance at the national and local levels. The massive migration flows from rural areas, which have largely contributed to population growth in most third world cities, indeed reflect government failure in agriculture, rural development and balanced national development. Even more obvious is the governmental failure regarding housing and living conditions in informal settlements and slums, where most newly arrived rural migrants are likely to settle and from where existing residents never seem to gain mobility to better neighborhoods.

It is not that national and urban governments of third world cities have not tried to tackle the population and housing problems. Indeed, there have been plenty of efforts, many of them put forward by international development agencies and their experts. National governments across Asia and Africa once implemented a very similar set of stick-and-carrot measures aimed at enforcing family planning, restricting rural–urban migration and eradicating urban slums, but few proved effective in stemming urban population growth. In fact, most third world cities experienced explosive population growth while their governments implemented these policies. Accusations abound, but national governments in the developing world are often singled out for blame, with serious questions raised about their willingness and capacity to gain effective control over urban populations in mega cities.

With the effectiveness and efficiency of those direct interventionist policies in question, some have argued for an accommodation-oriented approach to population growth in third world cities (Findley, 1993). Accommodationists first point out that explicit attempts to alter the migration patterns or growth rates of specific cities rarely achieve the intended result, and that rural migrants should be viewed as part of the solution, not the problem. They instead call for stronger policies to incorporate marginal populations of rural migrants and slum-dwellers into the city by fostering positive contributions of the informal sector and subsidizing the improvement of housing conditions in slums.

While many academics puzzle over population problems in the mega cities of poor countries (e.g., Tarver, 1994), others focus on emerging trends related to economic globalization. The assertion by Grant and Nijman (2002: 323) that "many cities in the less-developed world have moved through four historical phases: the pre-colonial, the colonial, the national and the global" rings particularly true for the large city regions, many of which have grown from colonial points of command to national economic centers and more recently to international gateways in their nations. Mega cities of the developing world attract national headquarters of both national and global companies. They are the sites of export-oriented production as well as platforms for the market penetration of foreign and transnational companies. When studying changes in Jakarta, Indonesia, researchers Dick and Rimmer (1998) point to the emerging urban forms, such as the gated communities of multinational real estate developers, as evidence of converging trends between Southeast Asian cities and North American ones. They do not deny the distinctiveness of Southeast Asian cities in the urban development history, but they do question labeling those cities as "third world." In the age of globalization, many third world cities increasingly gain elements that were previously exclusive to world cities.

There are exceptions to this general rule, though. Short (2004a), for example, identifies what he terms "black holes" and "loose connections" in the global urban networks. Some very large cities, such as Pyongyang in North Korea, remain only loosely connected to these trends of economic globalization. Still, in general, the mega cities of the developing world have become new production centers of the global economy and important sites for both cultural and economic globalization.

Primate cities

Mark Jefferson first identified the primate city in 1939. He argued that "in the early stages of a country's urban development, the city that emerges as larger than the rest develops an impetus for self-sustaining growth that enables it, over time, to attract economic and political functions to the extent that it dominates the national urban system" (quoted in Pacione, 2001: 79). The law of the primary city generally applies when the largest city houses a disproportionately large number of its nation's total population. A primate city, in common usage, refers to a city that accommodates more than twice the population of the next-largest city in the nation. The primary index of a city is calculated by dividing the population of the largest city by that of the second largest. Jefferson stressed that a primate city generally dominates the national economy of its country as much as its national urban hierarchy.

Primacy is not restricted to the developing world. Table 8.5 notes the existence of primate cities in both the developed and developing worlds. Paris and London, for example, are far and away the largest cities in their respective countries, both claiming primary indices greater than 7. Both cities dominate their respective countries: they hold the seats of government, they represent the centers of economic and social power, and they house the social elites as well as the vast

Table 8.5 Examples of primate cities

Country (year**)	Largest city (population, in millions)	Second-largest city (population, in millions)	Primacy index
Africa:			
Egypt	Cairo (6.8)	Alexandria (3.3)	2.1
Nigeria (1991)	Lagos (13.0)	Ibadan (3.0)	4.3
Kenya (1989)	Nairobi (1.3)	Mombasa (0.4)	3.2
Senegal (1999)*	Dakar (1.9)	Thies (0.3)	7.7
Asia:			
Iran (1996)	Tehran (6.8)	Mashhad (1.9)	3.6
Iraq (1987)	Baghdad (3.8)	Mosul (0.6)	6.3
South Korea (1995)	Seoul (10.2)	Pusan (3.8)	2.6
Thailand (1999)*	Bangkok (7.5)	Nanthaburi (0.4)	18.7
Eastern Europe:			
Bulgaria	Sofia (1.1)	Varna (0.3)	3.9
Hungary (1999)	Budapest (1.8)	Debrecen (0.2)	9.0
Latin America:			
Argentina (1991)*	Buenos Aires (11.2)	Córdoba (1.2)	9.3
Cuba (1998)	La Habana (2.1)	Santiago de Cuba (0.4)	5.2
Mexico (1990)*	Mexico, City (15.0)	Guadalajara (2.8)	5.3
Peru (1993)*	Lima (6.3)	Arequipa (0.6)	10.5
Developed World:			
France (1990)*	Paris (9.3)	Lyon (1.2)	7.7
Japan (2000)*	Tokyo (8.1)	Yokohama (3.4)	2.4
UK (1996)	London (7.0)	Birmingham (1.0)	7.0

Notes: * Population data are collected for the "urban agglomeration," while all other countries' data are for the "city proper"

** Population data for the latest available year

Source: Based on data from United Nations (2006)

majority of advanced producer services. The phrase "*Paris et le desert Français*" sums up the situation. The size of Paris and London in part reflects their colonial past: they served as imperial centers for empires that stretched around the world. While the colonies have disappeared, these urban centers retain their size and continue to dominate their national economies.

Arnold Linsky (1965) suggests that primacy is particularly evident in those countries with small areas, relatively large populations, low per capita incomes, export-oriented and/or agricultural economies, colonial histories and rapid rates of population growth. These conditions pervade the developing world. Not all developing countries, of course, exhibit urban primacy. China, Brazil and India, for example, are large countries with many large cities but not one dominant city.

Still, urban primacy remains a distinct feature of many national urban systems in the developing world. Table 8.5 lists examples from around the world, using the most recent United Nations data as a shared datum point (more recent data should be available on a case-by-case basis). Take the example of Mexico, where the urban primacy index of Mexico City is 5.3, as it is so much larger than the next-largest city, Guadalajara. Almost one in every five Mexicans lives in the Mexico City Metro region. Similarly, Lima, Peru, claims an index of 10.5. Out of a total population of over 25 million, almost 6.5 million live in the Lima Metro area. In Uruguay, one out of every two people lives in the primate city of Montevideo. In Thailand, the primacy is even more marked, with the dominant capital city of Bangkok (7.5 million) having more than 18 times the population of the next-largest city, Nanthaburi (481,000).

But what are the economic effects of urban primacy, both in the primate city itself and in the rest of the country? There are two competing schools of thought. Some argue for the notion of primate cities as generators of economic growth. While agriculture, which provides limited opportunities to the masses, dominates the rural areas, primate cities offer centers of economic and social opportunity. They house the more dynamic sectors of manufacturing, and more recently they have attracted service sector employment. As nodes of the global economy, they provide opportunities – the benefits of economic globalization. The wider provision of public service also provides a stronger social net for the weakest and poorest. This viewpoint sees the primate city as the outcome of agglomeration, as sheer size and activity pull additional residents to the city, and provide a platform for economic growth and social progress. Although Bangkok houses only about 14 percent of the country's population, it brings in 30 percent of the country's national income. Meanwhile, the extended metro region accounts for close to 40 percent of Thailand's GDP. Of the 11 major universities in the country, eight reside in the capital.

On the other hand, some see primate cities as a net drag on economic growth. The incredible concentration of people produces an exponential increase in negative externalities such as overcrowding, inefficient use of national space and underutilization of alternative urban areas. The pull of the primate city skews growth and development to only one city. The terms "overurbanization" and "hyperurbanization" embody this gloomy view of primate cities' effect on economic growth. This school of thought emphasizes the negative impacts of accelerated growth, including the exhaustion of environmental resources. In Bangkok, gross overcrowding of roads makes it common for people to spend up to three hours a day commuting. Air pollution and water pollution not only diminish the quality of life but seriously threaten health.

Case study 8.2

Third world cities in the league table of quality of life

Cities are rated for various attributes by various institutions. The indicators used in rating are as diverse as the cities. For American cities, *Places Rated Almanac* was a popular and useful reference in the 1980s and 1990s for those considering a move to another city. Now, the widespread use of internet research has prompted rating agencies to tailor ranking indicators to specific purposes, such as retirement, investment and even death – in effect, the quality of healthcare.

Some consulting agencies, such as the Economist Intelligence Unit (EIU) and Mercer Human Resource Consulting, now conduct annual surveys of urban living conditions on a global scale. Although international organizations such as the United Nations have made efforts to rank cities, not nations, in terms of slum, housing and informal sector jobs (UN-HABITAT, 2006–7), private companies evaluating a large number of international cities in terms of their quality of living remains a new phenomenon. The primary users of these survey reports are the multinational firms that regularly place employees on international assignments. In the survey, cities are evaluated and ranked on quality-of-living criteria that include political stability, crime, sanitation, air pollution, educational system, transportation and climate, among others. The EIU's global urban ranking is often used as a reference by multinational firms to assess "hardship allowance," or premium compensation paid to expatriates who would encounter harsh living conditions during international assignments.

As Table 8.6 shows, the world's most desirable places to live are all located in developed countries, particularly Northern Europe as well as North America (Canada in particular) and Oceania. In the meantime, most third world cities rank among the worst places to live and, therefore, places for expatriates to avoid at all costs. Baghdad remains the world's least appealing place for multinational firms' employees. The ratings for cities throughout Africa still rank very low, as they fill the bottom of the league table. The objectivity of the criteria used for ranking the cities can always be questioned. However, it should be noted that in the age of globalization, multinational firms still have a hard time finding expatriates willing to move to many third world cities.

Table 8.6 Most and least desirable cities for expatriates, 2006*

Rank	Top 20 cities	Country	Quality of living index	Rank	Bottom 20 cities	Country	Quality of living index
1	Zurich	Switzerland	108.2	196	Bamako	Mali	43.9
2	Geneva	Switzerland	108.1	197	Luanda	Angola	43.4
3	Vancouver	Canada	107.7	198	Tashkent	Uzbekistan	43.0
4	Vienna	Austria	107.5	199	Lagos	Nigeria	41.8
5	Auckland	New Zealand	107.3	200	Dhaka	Bangladesh	41.5
6	Düsseldorf	Germany	107.2	201	Conakry	Guinea	41.2
7	Frankfurt	Germany	107.0	202	Antananarivo	Madagascar	41.1
8	Munich	Germany	106.8	203	Niamey	Niger	41.1
9	Bern	Switzerland	106.5	204	Port au Prince	Haiti	41.1
10	Sydney	Australia	106.5	205	Kinshasa	Congo	40.7
11	Copenhagen	Denmark	106.2	206	Ouagadougou	Burkina Faso	40.5
12	Wellington	New Zealand	105.8	207	Nouakchott	Mauritania	38.2
13	Amsterdam	Netherlands	105.7	208	Port Harcourt	Nigeria	38.2
14	Brussels	Belgium	105.6	209	Sanaa	Yemen	38.2
15	Toronto	Canada	105.4	210	Ndjamena	Chad	37.2
16	Berlin	Germany	105.1	211	Pointe Noire	Congo	33.9
17	Melbourne	Australia	105.0	212	Khartoum	Sudan	31.7
18	Luxembourg	Luxembourg	104.8	213	Bangui	Central Africa	30.6
19	Ottawa	Canada	104.8	214	Brazzaville	Congo	30.3
20	Stockholm	Sweden	104.7	215	Baghdad	Iraq	14.5

Note: * Cities are ranked against New York City as the base city, which has a quality-of-living index score of 100. According to Mercer, 39 key quality-of-living determinants were used to calculate the indices for individual cities

Source: Mercer Human Resource Consulting

In an analysis of city ranking, Mercer's website notes the widening gap between the best and worst places: "In recent years, the gap between low-ranking and high-ranking cities has widened. While standards have improved in some regions, there remains a stark contrast between those cities where overall quality of living is good and those experiencing political and economic turmoil." International agencies such as UN-HABITAT (2003), as well as many private consulting firms, widely recognize the deterioration of living conditions in third world cities.

While economic opportunities and better access to public services attract migrants to primate cities, the high level of concentration produces negative externalities that may eventually outweigh the benefits of urban agglomerations. The long-term net outcomes will vary by city and by country. In some cities, urban planning measures, regional policies and centrifugal growth may lead to a geographical spread of economic activity and political influence. In others, the cities seem to suffocate by their own growth. At present we need more detailed studies of the costs and benefits of primate cities.

Further reading

Davis, Diane E., 2005, "Cities in global context: a brief intellectual history," *International Journal of Urban and Regional Research*, 29(1), pp. 92–109. This article critically reviews the current wave of writings on cities in the global context, as they seem to ignore the earlier generation of scholarship on third world cities that made considerable progress in examining cities in the global political and economic dynamics.

Drakakis-Smith, David, 2000, *Third World Cities*, London: Routledge. Drakakis-Smith made a significant contribution to the study of third world cities, particularly in Africa and Asia. This posthumously published book reviews development, population growth, employment, environment and planning and management issues of third world cities.

Robinson, Jennifer, 2006, *Ordinary Cities: Between Modernity and Development*, New York: Routledge. A very important book that challenges the established view of third world cities as an object for developmentalist intervention. While arguing for an ordinary-city approach in urban studies, Robinson seeks out more cosmopolitan resources for building a post-colonial understanding of cities.

Timberlake, Michael, ed., 1985, *Urbanization in the World-Economy*, New York: Academic Press. The third and fourth sections of this book examine regional implications and global patterns of urbanization in the context of the capitalist world system.

9 World city projects for national capitals

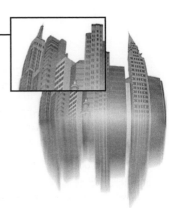

Learning objectives

- To think about the significance of capital cities in the national economic development of developing countries
- To understand national governments' efforts to boost the global status of their capitals
- To explore and compare different world city projects for third world cities

It is unfair and inaccurate to assume that, without any local agency, third world cities have been passively integrated into the capitalist world economy or linked into the global networks of production, financial capital and consumption. Evidence of global forces present themselves in every corner of those cities, whether through Nike's overseas contractors, local Western Union agents, internet cafés, International Monetary Fund resident representative offices or non-governmental organization training centers. Still, people, communities and governments in third world cities rework global forces on their own terms. In Chapter 6, we reviewed the ongoing debate on localizing forces in globalizing cities in the developed world. We now expand that debate to include those in third world cities.

In his ambitious project to synthesize the history of urbanization in China, Friedmann (2005) stresses the dual aspect of Chinese urbanization: the country's hyperurbanization began only a short while ago; nevertheless, Chinese cities stem from ancient origins. Given the recent and rapid integration of the Chinese economy into the world economy, one might point to global forces, such as trade and foreign investment, as the main causes of such radical change among large Chinese cities, particularly on the east coast. Friedmann, however, warns against this tempting interpretation, contending that "globalizing forces – economic, political and cultural – are accordingly viewed, not as the prime mover, but as complementary to and intersecting with an endogenous dynamic" (p. 449).

Among various forces comprising "an endogenous dynamic," this chapter focuses on national governments' efforts to improve their leading cities' international competitiveness. Governments in developing countries often take aggressive approaches to win international recognition for their national capitals' world city-ness or, at least, for their graduation from third world city status. Their world city endeavors include bidding to host the Olympics or other high-profile international sports competitions. Some undertake large-scale infrastructure upgrades, such as urban mass transit overhauls or new international airports. Some governments proudly announce plans to build the world's tallest building in the city center.

Since their independence, most developing countries have made efforts to modernize and, more recently, globalize their capital city as the centerpiece of their economic development plans. During the 1960s and 1970s, "developmental nationalism" materialized primarily in national capitals throughout the developing world. They have been designed to function as their respective countries' gateway cities to the wider world. Despite the myriad urban problems reviewed in the previous chapter, capital cities still symbolize national pride and express cultural identity. The first section of this chapter examines the development of capital cities in relation to various nation-building efforts made by national governments of developing countries. In the second section, we focus on the South Korean government's efforts to make its capital city, Seoul, world class.

National economic development and capital cities

Not all capital cities are the largest in their respective countries, but many are, especially in the developing world, where primate cities abound. Some capital cities represent former seats of centuries-old kingdoms, and therefore their urbanization roots pre-date colonial rule. In contrast, some capital cities grew dramatically as administrative centers of colonial governments. The historical processes of capital cities gaining economic dominance within their national urban hierarchies differ vastly, yet we attempt to find some common ground among large national capital cities in the developing world.

Peter Hall (1993) identifies seven types of capital city:

- Global capitals that perform supranational roles in politics and economics (London and Tokyo).
- Capitals that were created solely as seats of government yet often lack other functions, which remain in older, established and commercial cities (The Hague, Washington, DC, Ottawa and Brasilia).

- Former capital cities that have lost their role as seats of government but retain other historic functions (Berlin, Leningrad and Rio de Janeiro).
- Former imperial capitals that often still perform important commercial and cultural roles for the former imperial territories (Madrid, Lisbon and Vienna).
- Provincial capitals that once functioned as *de facto* capitals, sometimes on a shared basis, but have now lost that role, retaining, however, functions for their surrounding territories (Milan, Munich, Toronto and Sydney).
- Super-capitals that function as centers for international organizations (Brussels, Geneva, Rome and New York).
- Multi-function capitals that combine all or most of the highest national-level functions (London, Paris, Moscow and Tokyo).

Although Hall does not include any capital cities in the developing world as examples of the multi-function capital category, many of them indeed perform a combination of essential national functions in their respective countries. They serve as centers of national politics, economy and culture, and possess historical significance. For many capitals, the process of concentration and centralization began under colonial rule, as colonial governments established a number of port cities and administrative centers in Africa, Asia and Latin America. In the case of Latin America, for example, 18 of the present 20 capital cities were founded during the colonial period, and most of their populations grew dramatically from an influx of European immigrants during and after their independence (Hardoy, 1993). Bogotá, Buenos Aires, Lima, and Mexico City were the four viceroyalty capitals of the Spanish Empire during the nineteenth century.

Once independent, many of those former colonial capitals enjoyed inherited transportation and urban networks, allowing them to maintain a significant advantage over other domestic urban centers, although some countries, notably Brazil, Nigeria and Pakistan, launched new capital projects to negate this colonial legacy (Holston, 1989). Table 9.1 lists selected examples of new capital projects undertaken by newly independent governments. Some new capital projects symbolized the elimination of colonial legacy. Others aimed to reduce population pressures in former capitals so that their countries could achieve balance in terms of both population distribution and regional development.

It is clear that new governments across Africa, Asia and Latin America configured or reconfigured the internal structure of their capital cities, whether new or old, to meet the economic and political needs of newly independent nations. They adorned these redefined capitals with new national monuments and government buildings. In his seminal book on the historical geography of primate cities in Southeast Asia, McGee (1967) calls those cities "modern cult centres," acknowledging their roles as symbolic theatres of nationalism. Nationalism permeated most post-colonial societies in the 1950s and 1960s, and the political will to build

Table 9.1 New capital projects in the developing world since 1950

Year*	Country	Old capital	New capital
1956	Brazil	Rio de Janeiro	Brasilia
1957	Mauritania**	Saint Louis	Nouakchott
1959	Pakistan	Karachi	Islamabad
1961	Botswana**	Mafeking	Gaberone
1963	Libya***	Tripoli/Benghazi	Beida
1965	Malawi	Zomba	Lilongwe
1970	Belize	Belize City	Belmopan
1973	Tanzania	Dar Es Salaam	Dodoma
1975	Nigeria	Lagos	Abuja
1983	Côte d'Ivoire	Abidjan	Yamoussoukro
2005	Myanmar	Rangoon	Naypyidaw
2006	Palau	Koror	Melekeok

Notes: * The year when the new capital project was announce or approved
     ** Mauritania and Botswana did not have national capitals when they achieved independence but rather colonial administrative headquarters outside national borders
     *** Moved back to Tripoli in 1969
Source: adapted from Gilbert (1989: 235)

the "imagined community of a nation-state" (Anderson, 1983) inspired architecture and new civic design within capitals (Gottmann and Harper, 1990; Sutcliffe, 1993).

Besides being the political center of newly independent nations, many third world capital cities have served as catalysts of economic development. Almost all developing countries have channeled the vast majority of public investment into capital cities, leaving little for secondary cities and others further down the urban hierarchy (Rondinelli, 1983). This, in part, explains why the primacy of capital cities continued to rise in the second half of the twentieth century. Mega capital cities reflect disproportionate economic, political, cultural and educational investments made by national governments. As a result of the imbalance in resources and power, despite various national and international programs to tackle overpopulation problems, rural migrants flocked to the capitals. According to Richardson (1981 and 1987), in the 1970s, many governments in Africa and parts of Asia experimented with different population distribution policies, including forceful relocation, but such efforts conflicted with national economic planning policies that highly favored capital cities.

South Korea, a good example of an East Asian developmental state, fits this paradoxical equation perfectly. Its national government has enforced, since the

1960s, both decentralization policies on the capital city, Seoul, including a greenbelt to prevent the city's further expansion, and strategies for developing peripheral regions of the country. Some growth pole centers were established far away from Seoul to absorb rural migrants who otherwise would have headed for the capital city (see Case study 9.1). At the same time, the South Korean government continued economic development plans around Seoul, allocating a disproportionate share of the nation's infrastructural investment to the capital city and its vicinity. Despite decades of effort to reduce population pressure on Seoul, the metropolitan area still accounts for nearly half of South Korea's total population. This illustrates how the country's centralized economic development strategies dwarfed its policies for decentralization and redistribution of population. In the 1970s and 1980s, when South Korea was touted as a newly industrializing Asian economy, Seoul led the national economy in manufacturing production. The city's economic structure has since shifted toward the service sector, particularly business service, and Seoul's share of the national economy is higher than ever (Kim, 1998). Simply put, Seoul accounts for much of South Korea's economic success.

As national economic centers, capital cities are major destinations for foreign investment. Capitals and other big cities throughout the South are integrating increasingly into the global production and marketing networks of multinational corporations, although more than three-quarters of the world's foreign investments land in developed countries. Foreign companies tend to favor capital cities over other urban centers because of their better infrastructure, geographical proximity to government offices and larger pools of skilled labor. The growing presence of global companies in developing world capitals can affect their local property markets, which respond to the new demand with high-rise office buildings, five-star hotels, upscale residential areas and international airports.

In his research on foreign corporate activities in Ghana, Grant (2001) notes the uneven spatial effects of foreign investment on the national urban hierarchy. According to his survey, 84 percent of all foreign companies established in Ghana are headquartered in Accra, the capital city. Kumasi is the second-largest city, yet it hosts a mere 9 percent of all companies investing in Ghana. Grant points to the Ghanaian government's promotion of the capital city as an economic growth pole as the main reason for Accra's predominance in foreign company activities. This policy dates back to the early independence era, but similarly, for the past two decades under the liberalization regime, the government has given top priority to modernizing the physical infrastructure of Accra. Based on interviews with foreign companies, Grant concludes that access to the major international transportation hubs, both sea and air, as well as proximity to financial and governmental institutions, drove foreign companies to establish headquarters in Accra. The Ghanaian capital perfectly illustrates how the national economic dominance

Case study 9.1

Growth pole theory

French economist François Perroux introduced the concept of the growth pole, "*pôle de croissance*," in 1955. It was simply defined as a "propulsive" industry used to innovate and stimulate, as well as dominate, an array of linked industries. At the center of a dynamic and highly integrated set of industries, a growth pole should generate various economic benefits for participating firms through linkages and multipliers. Some of the benefits would include innovations and information-based/knowledge-based relationships as well as cost-based advantages (Chapman, 2005).

Initially an aspatial model conceived in abstract economic space, it became widely adopted during the 1950s and 1960s as a spatial strategy for stimulating economic development in peripheral or declining regions (Lasuen, 1969; Parr, 1999). Growth pole strategies were used by many governments at that time to promote manufacturing-led development, sectoral specialization and regional specialization, although their planning and implementation processes and outcomes varied widely by region. In the 1970s, however, as economic difficulties faced by many formerly specialized industrial cities mounted, the optimistic appeal of growth pole theory faded in regional planning and regional development policy circles.

In developing countries, growth pole strategies have often been associated with government efforts to correct overwhelming urban problems in primate cities. In order to decentralize the national urban system, national governments have channeled public investment into selected growth centers, located far from the primate, capital city. The development of Chiang Mai in northern Thailand illustrates how central governments design growth poles with the goal of reducing population pressure in the primate city of the country and, in broader terms, achieving a balanced national development. Along with decentralization, the spread effects and trickle-down effects of growth poles in the regional economy have been used to justify development policies that funnel public investment to selected growth centers, while leaving out the rest of the region and the nation.

Growth pole strategies have been criticized for creating acutely uneven development in developing countries that may already suffer from great inequalities among regions and between urban areas and rural areas, often originating in colonial rule and export-oriented economic structure. Instead of enjoying trickle-down effects from growth poles, many rural areas and small and medium-sized cities suffer from continued drains on investment capital, human capital and infrastructure.

of capital cities has accumulated over a long period of time, extending from pre-colonial, through colonial, to national and global eras.

Many capital cities in developing countries now boast some of the world's tallest buildings, often viewed as signs of confidence and optimism among national politicians, media and citizens alike. Anthony King (2004) makes a sophisticated analysis of the geographical shift of the world's tallest buildings from the US to other parts of the world. In the first half of the twentieth century, skyscrapers were identified solely with American modernity and corporate power. But in recent years, Asian cities (especially capitals) have competed to build the world's tallest building and thereby demonstrate their own modernity and economic success. King notes that "for reasons apparently mimetic of the United States, the spectacular high rise building has become a metaphor of modernity, if not worldwide, at least in some postcolonial or 'emerging' nation states" (p. 12). Building the world's tallest building has come to be associated more with capital cities in developing countries, particularly Asia.

Table 9.2 lists the world's twenty tallest buildings in 2006. Fourteen are located in Asia, only four in the US, one in the United Arab Emirates and one in Australia. China's southeast coastal cities, including Shanghai and Hong Kong, literally tower over the rest of the world, symbolizing the recent integration of China into the world economy and the economic prosperity it has brought to the region. Besides these Chinese cities, Malaysia's capital, Kuala Lumpur, houses several internationally recognizable buildings (Plate 9.1). Its Petronas Towers were the world's tallest until Taiwan's Taipei 101 went up in 2004. In addition to those twin towers, Menara Telekom ranks twenty-sixth highest in the world. The Petronas Towers were built as part of the "Vision 20/20," a national development plan to make Malaysia a developed nation by the year 2020. This nationwide mobilization campaign projected that Kuala Lumpur, its major beneficiary, will become a multimedia super-corridor and regional Islamic financial center, as well as a regional education center (Bunnell, 2004; Yap, 2004). The plan reinvented the capital city. In addition to those tall buildings, a new international airport, a new administration center with clustered government offices and high-tech parks all cropped up. Although the 1997 financial crisis sparked serious concerns about the effectiveness of the Vision 20/20 campaign, Kuala Lumpur's symbolic high-reaching makeover has attracted a great deal of international media attention.

Graduating from third world city status

The pursuit of world city status for capital cities stands out among political tactics to boost the global standing of a nation. No country has been more successful in

Table 9.2 The world's tallest buildings, 2006

Rank	Building	City	Year of construction
1	Taipei 101	Taipei, Taiwan	2004
2	Petronas Tower 1	Kuala Lumpur, Malaysia	1998
3	Petronas Tower 2	Kuala Lumpur, Malaysia	1998
4	Sears Tower	Chicago, USA	1974
5	Jim Mao Tower	Shanghai, China	1998
6	Two International Finance Centre	Hong Kong	2003
7	CITIC Plaza	Guangzhou, China	1997
8	Shun Hing Square	Shenzhen, China	1996
9	Empire State Building	New York City, USA	1931
10	Central Plaza	Hong Kong	1992
11	Bank of China Tower	Hong Kong	1990
12	Emirates Office Tower	Dubai, United Arab Emirates	2000
13	Tuntex Sky Tower	Kaohsiung, Taipei	1997
14	Aon Center	Chicago, USA	1973
15	The Center	Hong Kong	1998
16	John Hancock Center	Chicago, USA	1969
17	Shimao International Plaza	Shanghai, China	2006
18	Minsheng Bank Building	Wuhan, China	2006
19	Ryugyong Hotel	Pyungyang, North Korea	1992
20	Q1 Tower	Gold Coast City, Australia	2005

Source: Emporis Buildings (2006)

**Plate 9.1
The Petronas
Towers in
Kuala Lumpur**

Case study 9.2

Johannesburg: building an African world-class city

Johannesburg is often mistakenly thought of as the capital city of South Africa. In fact, the country has three official capitals: Bloemfontein, Cape Town and Pretoria. Johannesburg, the provincial capital of Gauteng, is, however, the most populous city of South Africa, with a population of 3.5 million. In addition, as shown in Chapter 5, it is identified by most world-cities researchers as *the* international gateway to and from sub-Saharan Africa (see Figure 5.1, above).

The city grew rapidly in the late nineteenth century when gold mines were discovered near by – Johannesburg's other name, iGoli, means "the place of gold" in Zulu. Along with European colonials, a large number of migrant workers arrived from the southern African region in the late nineteenth and early twentieth century. During the colonial and then apartheid periods, Johannesburg was marred by racially determined spatial segregation. The poor Africans lived densely in the south or on the peripheries of the far north of the city, while the rich Europeans lived in the suburbs of the center and north. The shanty towns of Soweto (South Western Townships) became a symbolic center for black resistance against apartheid injustice (Beavon, 1997; Smith, 1992). Throughout the apartheid era, poor and black residential areas suffered greatly from systematic underinvestment in infrastructure. The markedly uneven provision of public services across neighborhoods reflected a highly fragmented and racialized form of urban governance.

In 1994, apartheid ended and South Africa held its first non-racial election. Under the slogan of a "Unicity," the black townships have since integrated into a single, metropolitan-wide administration, named the City of Johannesburg Metropolitan Munici-pality. However, an outbreak of violent crime soon plagued the city center. Some inner-city neighborhoods, such as Hillbrow, have deteriorated badly since the mid-1990s. Crime prevention ranks as a high priority for the new government, in addition to building and upgrading infrastructure that could foster connections between formerly segregated townships (Robinson, 2006).

Amid social, political and infrastructural challenges, Johannesburg has adopted a set of initiatives that envision a better future (Beavon, 2006; Parnell, 2007). The Joburg 2030 Plan was launched in 2002 and aimed to "build an African world-class city" (Figure 9.1). Robinson (2006) notes that this slogan proclaiming an *African* world city reveals, on one hand, a collective awareness by city officials and elites of their city's African cultural heritage and close connections, through migration flows, to other parts of the continent. On the other hand, it may also reference the many difficulties in

Figure 9.1 Jo'burg: an African world city and a World Cup city

Source: **Official website of the City of Johannesburg, South Africa**

managing a city that is experiencing a combination of expanding slums and informal sectors, soaring crime, failing infrastructure and growing public health concerns.

Johannesburg faces daunting challenges, both to remedy the problems it has inherited from the past and, at the same time, to prevent future problems. Still, the city strives to transform itself into an African world city that competes against emergent world cities in the North. Under Jo'burg 2030, this former southern African mining town will emerge as a global center for financial and business services, information and communication technologies and business tourism. But in order to achieve this world-city status, notes the plan, Johannesburg must first tackle crime. South Africa will host the 2010 FIFA World Cup finals, with both the opening and final matches being held in Johannesburg. The city views this sporting event, the most globally watched media spectacle, as an opportunity to eclipse its old, mostly negative image and construct a new model of a booming, safe and multicultural city.

this than South Korea. Although the country was recognized for its phenomenal economic success as early as the 1970s, it still suffered from an image of poverty, tarnished further by the Korean War and the cold war. Yet, the 1980s brought dramatic change when Seoul was elected as the site of the 1988 Olympics. After successfully staging the Games, Seoul began to appear more frequently on maps of world cities and less on those of third world cities. This section examines how the South Korean government used international sporting events to sell its national capital's world city-ness.

First, a brief developmental history of Seoul. In 1394, Seoul, then called Hanyang, became the capital of Korea when the Yi dynasty relocated its palace there. The Japanese colonial government, when it established control over Korea in 1910, changed its name to Kyongsong. Following its 1945 liberation, the Korean state divided into North and South Korea. Seoul gained its current name in 1949 when it became the capital of the southern half. Although centrally located on the Korean peninsula, the city now found itself on the border between two hostile countries.

The country's first census reported slightly fewer than 1.5 million people living in Seoul in 1949. The dramatic growth of export-oriented manufacturing presented an opportunity for South Korea, which established a major proportion of new industry around Seoul. The city thus became a primary magnet for migrants from the rest of the country. Despite government efforts to curtail the flow of rural migration, Seoul's population grew rapidly until the 1980s and reached the 10-million mark in 1988 (Table 9.3). The city has accounted for well over 20 percent of the national population since the mid-1970s, while the second-largest city, Pusan, is home to less than half that figure. Seoul exemplifies the typical primate city, not just in population, but because of its leading role in politics, finance, business, higher education, culture and national identity.

Overall, Seoul's rapid population growth has slowed in recent years, but that has not been in the case in Outer Seoul, which consists of the city of Inchon and Kyonggi Province. Combined with Seoul itself, this is known locally as the Seoul Metropolitan Area or the Capital Region (Figure 9.2). Home to South Korea's largest airport and seaport, Inchon has long served as the gateway for Seoul's

Table 9.3 Population changes in Seoul, 1949–2005

Year	Seoul	Outer Seoul	Seoul Metropolitan Area	Share of national population (%)
1949	1,446,019	2,740,594	4,186,613	20.8
1955	1,574,868	2,363,660	3,938,528	18.3
1960	2,445,402	2,748,765	5,194,167	20.8
1965	3,793,280	3,102,325	6,895,605	23.6
1970	5,535,725	3,358,022	8,893,747	28.3
1975	6,889,502	4,039,132	10,928,634	31.5
1980	8,364,379	4,933,862	13,298,241	35.5
1985	9,639,110	6,181,046	15,820,156	39.1
1990	10,612,577	7,973,551	18,586,128	42.8
1995	10,231,217	9,957,929	20,189,146	45.3
2000	9,895,217	11,459,273	21,354,490	46.3
2005	9,762,546	12,858,686	22,621,232	48.1

Source: National Statistical Office, Republic of Korea

international trade. Seoul is literally surrounded by Kyonggi-Do, a province in which twenty-five cities house more than 100,000 residents each. Mounting mega city problems in Seoul have contributed to the recent growth of Outer Seoul, as government restrictions on manufacturing industries in the capital and the construction of large-scale apartment buildings in Kyonggi-Do have pushed and

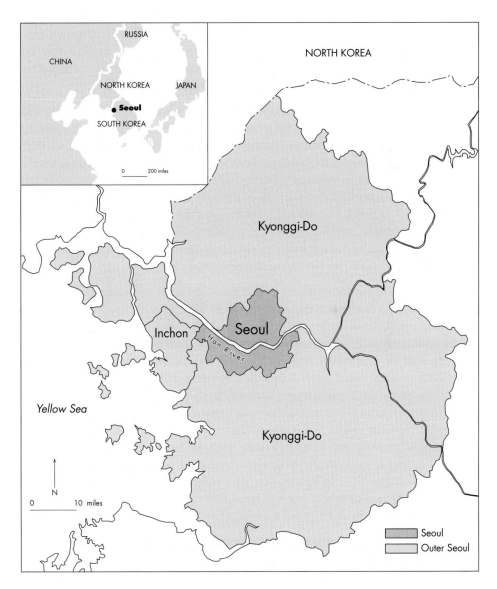

Figure 9.2 The Seoul Metropolitan Area

pulled residents and businesses out of the city center. The Seoul Metropolitan Area accounted for 48.1 percent of the national population in 2005, more than 22 million people – making it one of the largest cities in the world.

Although it had long served as capital, primate city and international gateway of South Korea, Seoul was not able to command much respect in the region until the mid-1980s. Tokyo stood at the apex of the Asian urban hierarchy, followed by Hong Kong and Singapore. Some Southeast Asian cities, like Bangkok and Manila, seemed to have more international connections than Seoul at that time. It was truly the city's success in hosting international sports and media events that prompted the global community to recognize Seoul as a world city, or at least as one of the most successful capital cities in the developing world.

The South Korean government considered an Olympics bid as early as the late 1970s, but the decision to "go for it" was made in 1980, while the country was still recovering from the latest of several military coups. Indeed, the new administration faced heavy criticism for bidding, since the odds seemed very much against it winning. Opponents repeatedly denounced the use of the Olympics to legitimize a military dictatorship. A survey following the Games, however, found that about 90 percent of South Koreans felt positive about the event (Kim *et al.*, 1989). It is clear that despite lingering worries about extreme actions from North Korea, heavy economic burdens and questionable political motives, the Seoul Olympics were a great success for South Koreans and their capital city.

It is difficult to quantify the economic gains and costs of hosting the Olympics: most sporting facilities are subsequently used for other purposes, and much of the human and indirect investment is not counted. Still, the Seoul Olympics generated a substantial profit of 497 million dollars, putting to rest the country's concerns about an Olympics debt. Alongside these financial gains, the Games benefited Seoul in four major ways. First, the Olympics and associated media events increased the city's international visibility from the moment it was declared host in 1981. Besides the standard obscurity that even major cities in the developing world face, the image of a war-torn country had continued to overshadow the dramatic progress that Seoul had made since the Korean War (Larson and Park, 1993). And as the capital city of a divided nation, Seoul had long been viewed through a filter of Cold War tension. The South Korean government effectively used the Olympics to reinvent its capital city as an emerging center in Asia and to enhance the country's global standing (*The Economist*, 2000).

The Seoul Olympics also allowed the South Korean government to discredit its northern opponent. The new message of economic success stood in sharp contrast to the economic difficulties that North Korea suffered. South Korea came to be viewed as a demonstration of the superiority of the capitalist system over its socialist competitor. Given the strained relationships between the two Koreas, concerns mounted over a potential Olympic boycott by the then socialist countries, but most of these nations officially recognized South Korea by sending their athletes to Seoul.

Third, preparations for the Olympics played a very important role in Seoul's urban redevelopment, especially on the Han River waterfront (Kim and Choe, 1997).. The city's mass transit system, including subways, was upgraded dramatically in the mid-1980s. A number of Olympic landmarks were built during this time, among them the Olympic Bridge, the 88 Freeway and the Olympic Park. The staging of the Games also provided a compelling rationale for the massive displacement of poor neighborhoods, many of which were transformed into business and financial districts.

Finally, the Olympics played a significant role in opening the eyes of the people of Seoul to the wider world. They tasted other cultures as they hosted 160 national teams and hundreds of thousands of tourists from many parts of the world. Seoul was not a cosmopolitan city and such exposure inspired increasing interest in the outside world in the years following the Olympics. The explosive growth of overseas travel since the late 1980s indicates how the Games fostered a growing sense of openness and globalness in Seoul.

Having made a success of the Olympics, South Korea applied for the 2002 World Cup finals and was awarded the right to co-host the event with Japan. It would have thrilled Koreans to beat Tokyo in the bidding process, but the Federation Internationale de Football Association (FIFA) named Seoul as the venue for the opening ceremony and Yokohama for the closing. The South Korean government viewed the World Cup as an unprecedented opportunity for Seoul to gain "legitimate" world city status by matching, if not outperforming, Tokyo, irrefutably a world city. During the years leading up to the event, the government made sure that its capital city could compete with Tokyo in such areas as its stadium, event organization, telecommunications equipment and even traffic management. However, many Koreans still worried that Tokyo would upstage Seoul in the international media, and concerns mounted as the event approached. But ultimately the Koreans exceeded their own goals and expectations. Thanks largely to South Korea's strong showing in the tournament (they unexpectedly reached the semifinal), Seoul received much international media attention. The red-clad South Korean fans, called the "Red Devils," created some of the most memorable images in the entire event.

Although it is still too early to gauge the effects of the 2002 World Cup on Seoul's world city status, the event certainly gave Koreans a level of confidence that they had lacked. It may be too ambitious for Seoul to expect an instant rise up the global urban hierarchy, yet the city has markedly improved its international standing by faring well in comparison with Tokyo, a preeminent world city, in hosting the most watched international sports competition. The 1988 Olympics was truly a landmark event for Seoul's international visibility, and the 2002 World Cup reinforced that perception of Seoul as a world city rather than a third world city.

After Seoul demonstrated the value of the Olympics as an avenue towards world city states, many other capital cities in the developing world have applied to host the Games. Table 9.4 lists the Olympic host cities and the bid cities that have made the final round since 1980. When Beijing stages the 2008 Games, it will be the third Olympic city of the developing world, after Mexico City in 1968 and Seoul in 1988. Both Belgrade and Istanbul made strong cases to host the event, while Buenos Aires and Cape Town were among the five final candidate cities to host the 2004 Olympics. Although not listed in the table, many more third world cities have made it through to the runner-up rounds in the recent past, including Bangkok, Cairo, Havana, Kuala Lumpur, Rio de Janeiro, St. Petersburg and San Juan.

Hosting the Olympics and other world-class events creates a global stage on which a city or a nation can promote its economic and cultural achievements. Once a city has established a positive image, policies designed to attract investment and tourists have better prospects for success, and world city status may come within reach. However, this should not be interpreted as an argument that an Olympic bid represents the best practice for effective governance in third world cities. We have emphasized in Part Two that there is no such thing as the "best practice" in urban governance, although some politicians pretend otherwise. The strategy that

Table 9.4 Host and bid cities of the Summer Olympics, 1980–2012

Year	Host city	Bid cities (finalists)
1980	Moscow	Los Angeles
1984	Los Angeles	–
1988	Seoul	Nagoya
1992	Barcelona	Amsterdam, Belgrade, Birmingham, Brisbane, Paris
1996	Atlanta	Athens, Belgrade, Manchester, Melbourne, Toronto
2000	Sydney	Beijing, Berlin, Istanbul, Manchester
2004	Athens	Buenos Aires, Cape Town, Rome, Stockholm
2008	Beijing	Istanbul, Osaka, Paris, Toronto
2012	London	Madrid, Moscow, New York, Paris

Source: International Olympic Committee (2006)

the South Korean government used to court the coveted world city status may remain beyond the reach of, or prove ineffective for, most other developing countries.

Further reading

Gugler, Josef, ed., 2004, *World Cities beyond the West: Globalization, Development and Inequality*, Cambridge: Cambridge University Press. This edited collection brings together some fascinating empirical studies of the impact of globalization on non-Western world cities where diverse political and public discussions take place concerning their world city status.

Richardson, Harry W., 1981, "National urban development strategies in developing countries," *Urban Studies*, 18 (3), pp. 267–283; Richardson, Harry W., 1987, "Whither national urban policy in developing countries?," *Urban Studies*, 24, pp. 227–244. These two articles critically assess various national urban development policies which have been designed to slow down the excessive growth rates of capital cities in the developing world. The second looks at the failure of such policies in developing countries.

Segbers, Klaus, ed., 2007, *The Making of Global City Regions: Johannesburg, Mumbai/Bombay, São Paulo, and Shanghai*, Baltimore: Johns Hopkins University Press. Four of the largest and fastest-growing urban regions in the developing world are chosen for a comparative study of the political making of global cities. The contributors examine the differences, as well as commonality, between these four "southern contenders competing with the northern global cities." The final chapter of the book contains comments made by policy-makers of those four cities.

Taylor, John, Jean G. Lengellé and Caroline Andrew, eds, 1993, *Capital Cities: International Perspectives*, Ottawa: Carleton University Press. A collection of twenty-four articles on the definition, role, symbolism and future of national capital cities across the world.

10 Globalizing islands in developing countries

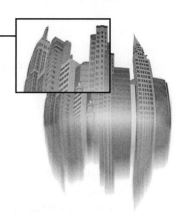

Learning objectives

- To have a critical understanding of the uneven nature of globalization
- To understand growing inequality and disjuncture within third world cities
- To consider the concept of splintering urbanism

Based on his students' definitions of globalization, Murray (2006: 3) describes a common image of globalization as "a process that unfolds like a blanket across the globe, homogenizing the world's economies, societies and cultures as it falls. Everywhere becomes the same, boundaries don't matter and distance disappears." In stark contrast, inequality among and within countries often accompanies globalization. This globalization-induced divide often justifies the view that the developing world is worse off in the wake of global economic progress (Sachs, 2005). This chapter focuses more closely on growing inequalities within developing countries and particularly in large cities.

As shown in Chapter 5, world cities illustrate the widening gap between the winners and losers of globalization. They house a growing number of both highly paid "command and control" jobs for corporate executives and low-wage "cleaning" jobs for immigrant workers (Sassen, 1991). The expanding enclaves of both poverty and wealth in world cities underscore the uneven nature of globalization processes. Third world cities also experience the widening gap between the winners and losers of globalization. While the few with ties to global corporate services and consumer cultures have emerged as the new rich in many third world cities, the many without such ties have suffered. The relationship between globalization and inequality emerges most vividly in world-class networked infrastructures and gated communities in third world cities. Such features further marginalize the vast majority of city residents (Graham and Marvin, 2001; Mycoo, 2006).

The contradictory and uneven nature of globalization can also be seen among different spheres of urban change in third world cities. People, money, goods, policies, jobs and technologies within a city have traveled along different paths and at different speeds in linking to the outside world. As we saw in the previous chapter, the South Korean government has eagerly pursued the globalization of its capital city, Seoul, but it also demonstrates unmistakable reluctance in accepting and adapting to the growing arrival of migrant workers from the rest of Asia, including China and Southeast Asian countries. As a result, Seoul exhibits all the important features of an economically globalized city but fails to foster ethnic diversity, another important characteristic of world city-ness. This chapter examines growing inequalities and disjunctures in globalizing third world cities, with particular focus on Bangalore in India and Seoul.

Inequality in globalizing third world cities

Urban inequality pre-dates the recent history of global urbanism. Since the dawn of cities, certain groups of urban residents have suffered limited access to the socio-cultural, economic and political opportunities readily available to others, although the nature and extent of exclusion differ vastly among places and over time. Along with poverty, inequality represents a main characteristic of third world cities – although Sassen (1991) finds a strong association between global city-ness and social polarization also in emergent world cities like London and New York. One archetypal image of third world cities portrays a minority of extremely rich elites living alongside the majority of impoverished slum residents with no concern for their suffering. Global news media help to perpetuate this negative image of third world cities. In fact, international media rarely portray such cities in a positive light. However, it should also be noted that poverty issues in third world cities can hardly be tackled without the privileged elites willing to change their role in the developing society.

As mentioned in Chapter 8, the already wide gap between the urban rich and the urban poor in third world cities has increased in recent years. Their respective national governments have received most of the blame for the growing inequality, as they have failed to show genuine, sustained political will and capacity to tackle the root causes of the problem. Paradoxically, international development agencies, such as the International Monetary Fund, have contributed to the rising disparity in third world cities, as their neo-liberal policies and privatization programs leave a large number of urban residents without basic public services (UN-HABITAT, 2003: 6). This section looks into the effects of the recent globalization process on such growing inequality, as many developing countries, and particularly their largest cities, have increasingly integrated into a global economy in the past decade or so. Put simply, the existing inequality in many third world cities has worsened as the few rich benefit from their countries' growing engagement with the global networks of firms and value chains, while the many poor suffer increased poverty by the same process.

In their groundbreaking book on the relationships between cities and networked technologies, *Splintering Urbanism*, Graham and Marvin (2001) point to the building of modern and globally connected infrastructure as a major source of growing inequalities in third world cities. For decades now, note Graham and Marvin, modernized networks of transportation and telecommunications infrastructure have been seen as clear visual evidence of economic development and progress. Development policy-makers and academics widely, if not universally, accept the view that any responsible, forward-looking government should make significant investments in infrastructure. So much so that the frenzy of 1990s urban mega projects in large cities in the developing world were justified as the provision of standard public goods (Siemiatycki, 2006; Yusuf and Nabeshima, 2006). As the concept of urban competitiveness became a dominant discourse within public policy circles, an even greater emphasis was placed on upgrading telecommunications infrastructure as a strategy for competitiveness in the global market.

However, Graham and Marvin challenge this widely accepted view by asking whether the construction of new state-of-the-art international airports or wireless telecommunications infrastructure truly serves the urban general public, as governments routinely claim. They argue that such urban infrastructure efforts help create urban spaces for the socio-economic elite rather than the whole population. In other words, despite the great deal of time and public money customarily spent on networked infrastructures, such efforts are likely to splinter, not integrate, various factions of the city.

Graham and Marvin identify the origin of splintering urbanism in Western cities during the period between 1850 and 1960 when the provision of networked infrastructures became gradually centralized and standardized. During the same time, the West imposed the ideal of a standardized infrastructure network upon the rest of the world, including many colonies and, later, newly independent developing countries in Africa, Asia and Latin America. While upgrading transportation and telecommunications infrastructure served the need of colonizers and their local associates during the colonial period, today's modern and postmodern infrastructure networks meet the demands mostly of transnational elites, such as multinational firms' local operations, globally connected local firms and young educated people.

This process of infrastructural dualization repeats in today's growing digital divide in third world cities. This refers to a gap in opportunities among different populations and different places to access advanced information and communication technologies. Despite the popular perception of a phenomenal increase in the use of computers and the internet these days, a vast number of people and neighborhoods in third world cities remain untouched by computer networks and other advanced information and communication technologies.

The growing digital divide is often addressed on the international scale, as many developing nations cannot afford the high-performance and high-speed infrastructure readily available in developed nations (UNCTAD, 2005). In the past decade, however, almost all countries around the world, at least within their large cities, have been connected to a global network of fiber-optic cables and satellites. A less palpable but more significant divide is now observed within third world cities where expensive, publicly funded network infrastructure serves only a few haves, excluding the many have-nots. Though an essential element in new urban landscapes of innovation, economic progress and cultural transformation, globally networked infrastructure intensifies socio-economic inequalities. The availability of internet connections varies greatly across neighborhoods within cities. In the case of Chennai (formerly known as Madras), the fourth-largest city of India, for example, upper-class residential areas, such as Boat Club Road, have enjoyed high-speed connections for years, while most other neighborhoods in the city struggle with shortages of electric power, let alone internet provision.

Splintering urbanism takes a turn for the worse when these networked infrastructures are not locally operated or regulated. In fact, most of them conform to the standard set by global networks. Those using the network effectively detach from local practices and norms in connecting to global circuits of economic and technological exchange. This is not to say that the national or local governments of those third world cities should stop upgrading transportation and telecommunications infrastructure. Instead, they could work to expand the benefits of the modern infrastructure to the general population and, at the same time, to pass on a significant portion of the infrastructural cost to the super-connected people, firms and institutions – in the case of Boat Club Road in Chennai, mostly foreign expatriate families.

Besides the digital divide, income inequality has also been growing in many third world cities that have integrated into the global network of firms, banks, services and technologies. In his research on the corrosive effects of globalization on social integration in Istanbul, Keyder (2005) points to the rise of the new rich, including financial analysts, software programmers and other young professionals, who have close ties to global corporate services and consumption patterns. While economic globalization has profited many young professionals and bankers, notes Keyder, the vast majority of city residents suffer from new levels of inequality and polarization in employment, income and the use of the built environment. He goes on to say that "within the global ideological climate of neoliberalism these developments [of inequality and polarization] are not sufficiently counterweighted by social policy" (p. 124).

The growing concentration of income and wealth has manifested in gated communities that have been springing up in many large, globalizing cities of the developing world. Rich neighborhoods wall themselves away from neighboring residential areas in an effort to improve security, in what is often called the "architecture of fear." In addition to physical designs of restricted access such as walls, gates and armed security guards, gated communities are often characterized by their private provision of public goods and services. Although luxury residential enclaves have existed in third world cities for centuries, gated communities seemed to gain unprecedented popularity among the upper middle class in the last two decades of the twentieth century (Webster *et al.*, 2002). Scores of upper- and upper-middle-class gated communities have developed in cities of east coast China (Miao, 2003; Wu, 2003), Latin America, the Caribbean (Mycoo, 2006; Roberts, 2005) and Southeast Asia (Dick and Rimmer, 1998) as well as in Arab and South African cities (Murray, 2004).

Those gated communities transform the overall landscape of their cities, which witness unprecedented real estate booms for high-end housing markets that accommodate young, high-income and more Westernized professionals. Instead of closing the growing divide between the few rich and the many poor, governments often facilitate the creation of fortified enclaves by privatizing public land, relaxing regulations on real estate development and providing transportation and telecommunications infrastructure. The fear of crime and violence among the rich could also be traced to the failure of governments effectively to police the city, although gated residential areas are as much the product of perceived threat as reality.

Gated communities, much like slums, are therefore a socio-spatial indication of growing inequality in third world cities. In Mumbai 4 million people are believed to be living in the city's slums. One UN-HABITAT report (2003) notes that the global number of slum-dwellers increased by about 35 percent during the 1990s, and in the next three decades it will double to 2 billion if no concerted action is taken globally to address the problem. Mike Davis (2006: 200), who earlier warned against Los Angeles' growing socio-economic polarization in *City of Quartz* (1990), sums up his concern about the widening gap in third world cities and the growth of slums: "[I]f informal urbanism becomes a dead-end street, won't the poor revolt? Aren't the great slums just volcanoes waiting to erupt?"

Case study 10.1

Gated communities in Trinidad

Trinidad is the larger and more populous of the two main islands of the Republic of Trinidad and Tobago, a Caribbean country rich in oil and natural gas. Home to the vast majority of the country's 1.3 million population, Trinidad houses all its major cities, including Chaguanas, San Fernando and Port of Spain, in the west. Residential segregation between different classes and different ethnicities has existed in these cities for centuries, yet the booming expansion of gated communities is a fairly new phenomenon to local people. Their construction began around the capital city of Port of Spain in the late 1990s, but their numbers have escalated dramatically since 2002. Now well over 200 gated communities stand in the western part of Trinidad, accommodating local professionals and transnational elites, most of whom are associated, directly or indirectly, with either oil and natural gas or offshore financial services.

Mycoo (2006) attributes the recent rise of protected suburban enclaves to state failure in providing necessary public goods and services, such as reliable water service and crime control. Responding to the failing public sector, the upper- and upper-middle-class households have established gated communities equipped with privately managed, higher-quality residential services.

Such claims naturally spark inquiry into recent government failures. Trinidad had an excessively large public sector in previous decades, Mycoo says. The Black Power riots of 1970 prompted the government to establish and expand these projects for the underprivileged. Thanks to the massive surges of oil prices in the 1970s, the government could afford to fund various subsidies in education, housing, health and water provision. Like many other Latin American and Caribbean governments at that time, however, the Trinidadian government engaged in excessive interventions and expensive programs. With oil prices falling in the mid-1980s, the country encountered a debt crisis, followed by structural adjustment programs mandating significant cuts in the public sector. Although Trinidad bounced back from the decade-long economic setback after the mid-1990s, unemployment rose steeply.

While low-income households suffer record highs for unemployment and poverty, Mycoo notes that public expenditure on the social sector remains low in the post-structural adjustment era. Consequently, discontent has grown in the area of social and public services. The unemployed and poor have demanded more and better government assistance, and their protests have grown increasingly violent and widespread. Based on a survey of 250 households in gated communities across the island, Mycoo concludes that rampant and serious crime, including kidnappings of wealthy individuals, combined with the government's inability to curb it, have driven the rich into the gated communities in the suburban areas of the Capital Region. The rise of these communities suggests state failure not only in providing the poor and unemployed with basic public services but in creating a social environment in which the rich can feel safe, or at least not threatened.

High-tech enclaves in Bangalore

Innovations in information and communications technologies remain the most important factor in the globalization process of economic activity. By lowering the cost and shortening the time needed to connect, people and companies in different time zones can work on the same project at the same time. American airline companies have long located their call centers in low-wage, English-speaking countries, such as India and Jamaica. Financial firms followed suit by outsourcing their human resource and customer service divisions overseas. Multinational corporations now consider outsourcing operations as a standard and essential cost-cutting measure. Cities or islands in the developing world have been labeled as "overseas call centers," "offshore financial centers," "offshore software development centers" or "IT outsourcing centers."

It is interesting to look at the socio-economic effects of globalization in these offshore centers, as they are connected to the core of the world economy only in a certain way and to a limited extent. Put simply, they service multinational corporations, often headquartered in world cities, in the areas of data analysis, customer services and software development. As these outsourcing centers are incorporated into the global economy, the gap between them and their non-connected populace widens. This section examines the growth of Bangalore as a global information technologies (IT) center and the various urban problems that accompany this growth.

The rapid growth of India's IT sector represents one of the two most celebrated success stories in the globalizing South (China's industrialization and export growth being the other). India's success in IT has been credited to a large pool of young, talented and English-speaking computer engineers (Plate 10.1). Parthasarathy (2004) stresses specific conditions prevalent in the 1990s that helped the Indian IT industry take off and continue to grow. They included an explosion in global demand for high-skilled, low-wage software professionals and the Indian state's neo-liberal policy initiatives, including the establishment of software technology parks, which are similar to export processing zones but

**Plate 10.1
Learn to
develop new
software in
Chennai, India**

dedicated to the software industry, in Bangalore. In addition, the time difference of 12.5 hours between Silicon Valley and India benefits Indian firms in bidding for outsourcing contracts since they can undertake offshore maintenance and re-engineering after American customers leave their places of work.

By 2000, India had become the largest software exporter among non-OECD countries, with its exports jumping from virtually zero in 1985 to $4 billion in 2000. Software revenues now account for almost 20 percent of the country's total export earnings. Bangalore (called Bengaluru in the local Kannada language) makes up around a third of India's software exports, followed at a distance by other large cities such as New Delhi, Chennai and Hyderbad (Parthasarathy, 2004).

The labels "Asia's Silicon Valley" and "India's IT capital" originated as a result of the success of home-grown software services giants (e.g., Wipro, Infosys Technologies and Tata Consultancy Services) and because of the presence of numerous IT multinational firms, including Oracle, IBM and Texas Instruments (Hamm, 2007). In addition, Reuters, the world's largest international multimedia news agency, has operated one of its four global data operations facilities in Bangalore since 2004 (Tiverton in the UK, New York and Singapore being the other three). According to Reuters' website, the Bangalore unit is responsible not only for processing financial data on global companies but for producing corporate earnings reports and broker research on US companies. In other words, Indian financial journalists in Bangalore cover American financial news.

The prosperous software industry has transformed this southern Indian city into one of the world's most famous hi-tech metropolises. Bangalore has become a center of profit and consumption in India. Software programmers have emerged as the new rich in a society that traditionally associates wealth with inheritance. The city's IT workforce is estimated at half a million, including non-resident Indians and expatriates, as well as locally trained computer engineers.

However, this newfound affluence has not benefited the rest of the city or the country in general (Keniston and Kuman, 2004). Rosenberg (2002) refers to the wealthy software companies and their engineers as a silicon island in a third world sea. According to the UN-HABITAT report (2003), three-quarters of India's urban households still live in slums, and they suffer from limited access to sanitation and clean drinking water. An article in *The Economist* (2006) reports on the recent rise of poverty rates among urban Muslims in India. On the one hand, says the article, the country prospers with its strong software industry; on the other, its 150 million Muslims fall further behind as their marginalization from basic public services, such as electricity, roads and water, continues. Davis (2006: 171) describes the growing divide caused by the IT success as follows: "Growth has been stupendously lopsided, with enormous speculative investment in the information technology sector leaving agriculture to stagnate and infrastructure to decay."

The acute disparity between Bangalore and the rest of India often mirrors inequalities inside the city itself. Although Bangalore has been praised as one of the world's hottest IT centers, computer engineers make up only a tiny proportion of the city's total population of 8 million. The sobering reality is that the vast majority of Bangalore's residents are not connected to global fiber-optic networks, nor do they have the necessary skills to use them. The IT companies and their engineers in Bangalore are indeed "islands of prosperity surrounded by an ocean of poverty." Local newspapers already talk about Bangalore's equivalent of the "Dutch Disease," a phrase that implies prosperity in one sector can hurt the rest of the economy (Shenoy, 2006). The IT success has brought Bangalore an unprecedented construction boom for building office parks, residential subdivisions, hotels, commercial strips and shopping malls. The spatial restructuring that the city has undergone in the past decade serves the needs of computer engineers, not those of the urban poor, who have been further marginalized by rising rents and utility costs. While a significant portion of the city's public funds have been spent on improving bandwidth and telecommunications infrastructure to please grumbling IT companies, the vast majority of residents helplessly watch as basic public services deteriorate. In his article on Bangalore as a divided city, Benjamin (2000) estimates that more than 40 percent of the city's population lives in slums.

The poor are not the only group to criticize city managers and policy-makers for Bangalore's socio-economic disparities. Some high-profile IT companies, presumably the beneficiaries of recent economic and spatial changes, have threatened to relocate their local operations unless the city's transportation infrastructure improves significantly (Moreau and Mazumdar, 2004). Bangalore's public infrastructure has lagged woefully behind the pace of private sector investment and population growth. The city's population has doubled since 1990, with the IT boom attracting not only college graduates with engineering degrees but a large number of rural migrants from across the state of Karnataka and south India in general. Its main roads exhibit a wide range of transportation methods, including passenger cars, three-wheeled auto rickshaws, scooters and animal-drawn wagons. Bangalore's struggle with poor infrastructure has been figuratively illustrated in global news media, including the BBC (2007) and *Newsweek* (Moreau and Mazumdar, 2004), with maddening traffic jams, pothole-filled roads and sewer overflows. In an annual survey of worldwide expatriates on the quality of life in different cities (Mercer Human Resource Consulting, 2007), Bangalore is ranked the 153rd best place to live among 215 cities, much lower than its Chinese rivals Shanghai (103rd) and Beijing (122nd). Its infrastructural woes could eventually blunt its competitive edge against other Indian cities and emerging IT centers across the developing world in attracting IT-related investment and outsourcing.

There is no debate that Bangalore has achieved truly phenomenal success in the past decade. Considering the economic crisis that India sustained for decades before its 1991 liberalization, the city's rise in the global software industry gives great hope to other parts of Indian society. It remains to be seen, however, whether Bangalore will truly transform itself into India's Silicon Valley, rather than Silicon Valley's India (Parthasarathy, 2004). While the former implies an IT center capable of defining and developing new products and technologies in the industry, the latter refers to a cluster of low-wage, back-office service firms tied to larger central operations in Silicon Valley, California. Either way, the continued success of IT companies and engineers will hinge on the nature and extent of the connections they develop not only with their customers in Silicon Valley but with their fellow-citizens in Bangalore.

Ethnic enclaves in Seoul

The uneven and irregular nature of globalization can be seen among different spheres of urban change in large cities of the developing world. People, money, goods, policies, jobs and technologies within a city have traveled along different paths and at different speeds in linking to the outside world. Appadurai (1996) conceptualizes the present global condition in terms of constant contradictions and disjunctures among the economic, political and cultural facets of globalization. This section looks at Seoul as an example of the disjuncture between

Case study 10.2

Bangalore: India's Silicon Valley or Silicon Valley's India?

India's computer software industry has grown so quickly that frequent comparisons have been made between Bangalore and Silicon Valley, unquestionably *the* center of the world's information technology industry. Bangalore has often been referred to as the "Next Silicon Valley" as well as "Asia's [or India's] Silicon Valley." Some go as far as to argue that Bangalore will eventually overtake Silicon Valley to become the world's premier IT center (Hamm, 2007). While Bangalore's software industry has been established around low-cost outsourcing services, international news media have noticed some early signs of a shift toward high-value work in this bustling Indian city. A recent *New York Times* article (Rai, 2006) notes that "the new firms are drawn by the [Bangalore] region's big pool of engineering graduates, many of whom have expertise in esoteric new technologies. That advantage, coupled with labor costs much lower than those of Silicon Valley, is starting to turn Bangalore, long a center for lower-end outsourcing services, into a center of higher-end innovation."

In his comparative study of Bangalore and Silicon Valley, Parthasarathy (2004) agrees that Bangalore has recently changed from a low-wage backwater into an important production center that develops products for the global software industry. However, he argues that the similarity between the two IT centers stops there, as India's rather small domestic software consumption limits Bangalore's further transformation into a region that defines new products and technologies on a consistent basis, as Silicon Valley does. He concludes that while Bangalore can be called Silicon Valley's India, it is constrained from becoming India's Silicon Valley.

Assuming that Bangalore can never develop into a global innovation center seems as premature as praising it as an imminent threat to Silicon Valley's leadership in the IT sector. However, Parthasarathy's term "Silicon Valley's India" illuminates the strong presence of Indian engineers in Silicon Valley as well as Bangalore-based outsourcing firms that serve IT firms in the valley. It is well known that Silicon Valley has long drawn top-notch engineers from abroad, particularly Asia. Foreign-born talent, according to the latest statistics (Joint Venture, 2007), accounts for 55 percent of the region's science and engineering workforce. Less is known about the Indian IT workers, who make up almost 14 percent of the region's total and a quarter of its foreign-born workforce (Figure 10.1). The Indian engineers working in Silicon Valley have formed professional associations, such as the Silicon Valley Indian Professionals Association, to provide opportunities for networking and information sharing as well as financial sources for ethnic entrepreneurs (Saxenian, 2004).

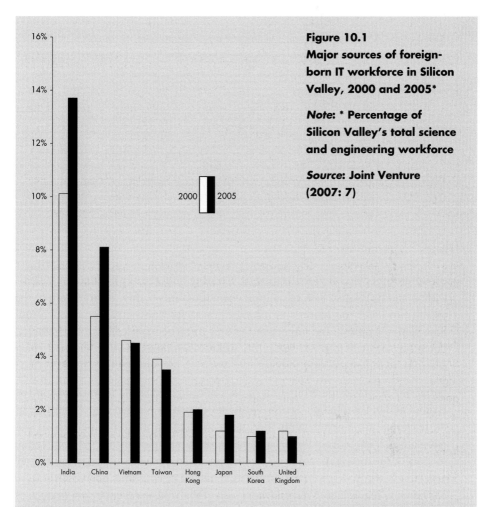

**Figure 10.1
Major sources of foreign-
born IT workforce in Silicon
Valley, 2000 and 2005***

Note: * **Percentage of
Silicon Valley's total science
and engineering workforce**

Source: **Joint Venture
(2007: 7)**

The strong presence of Indian engineers in Silicon Valley is often viewed as an indication of an Indian "brain drain," since a large portion of the best and brightest have left India to work for American companies. Based on her decade-long research on the impact of immigrant engineers and entrepreneurs in Silicon Valley, however, Saxenian (2006) argues that these foreign-born talents have played a key role in the recent development of IT industries in Asia, including India. They have not only connected US firms with India's low-cost, high-quality skills but have transferred cutting-edge technologies and market information to numerous startups in their home country. More recently, an increasing number of Indian professionals, after working in Silicon Valley for several years, now return home to start their own IT service businesses in Bangalore, positioning themselves at the center of the growing co-operation between the US and India in high-tech areas. Whether their return is temporary or permanent, Saxenian notes that those Indian engineers have helped develop transnational networks between Bangalore and Silicon Valley, and that a process of "brain circulation," rather than "brain drain," has taken place between the two.

Saxenian goes on to argue that the Indian IT talents in Silicon Valley could play a critical role in taking Bangalore to the next level, if the Indian government creates suitable incentives for them to return home and serve as policy advisors, investors, entrepreneurs and managers. If Saxenian's policy advice materializes, it might not matter whether Bangalore becomes India's Silicon Valley or Silicon Valley's India. Bangalore will simply become part of Silicon Valley and vice versa.

economic and cultural globalization, illustrating both the city's economic globalization and its relative lack of cultural cosmopolitanism, due, in large part, to the absence of labor migration and immigration in its world-city formation. Seoul now demonstrates all the important features of an economically globalized city, but it has not yet fostered the multicultural cosmopolitanism that also characterizes world city-ness.

Seoul has achieved the reputation of an emerging economic powerhouse in the Asian Pacific, yet in ethnic terms it remains a strongly Korean city. Almost all of its residents are Korean, with foreigners making up a mere 1 percent of the population. Viewed from the perspective of economic globalization, Seoul, home to many large multinational firms, banks and related business services, is clearly a world-class city (Kim, 2004). In addition, the city's consumer and youth culture has become remarkably Westernized, if not globalized, in recent years. Images of Seoul featuring postmodern-looking buildings, high-end stores and night-life scenes now greatly resemble those of a culturally globalized place. However, lacking a sizable population of immigrants, it has not cultivated a cosmopolitan demographic. It is argued that the lack of ethnic diversity in Seoul reflects its government's restrictive policies on immigration and labor migration.

Seoul has long been a major source of international migration flows. The population of ethnic Koreans residing overseas is estimated at 5.5 million (Overseas Korean Foundation, 2006). The vast majority of Korean emigrants or ethnic Koreans living abroad reside in China, the US, Japan and Russia – approximately 2 million ethnic Koreans are in China, another 2 million are in the United States, 1 million are in Japan and half a million are in Russia. The large Korean diaspora is a product of the economic and political difficulties that Koreans endured for much of the twentieth century, including Japanese colonialism and the Korean War (1950–3), followed by dire poverty.

In contrast to the large number of ethnic Koreans overseas, the number of foreign nationals residing in South Korea is a mere half million, including some 200,000 illegal migrant workers. Foreigners make up about 1 percent of the current total population of 47 million. In the 2005 census, the number of naturalized citizens was considered too minimal to be counted separately, although the South Korean government plans to include this statistic in the next census, scheduled for 2010. While Seoul started globalizing its economic and political relations immediately after the 1988 Olympics, the city's ethnic composition continues to be Korean. It is one of the very few large cities in Asia that did not develop an economically vibrant overseas Chinese community in the 1950s and 1960s; and it was virtually bypassed by Vietnamese refugees in the 1970s, as the South Korean government staunchly refused to offer them assistance. Until recently, Seoul was also not a major destination for overseas Filipino workers, whose migration flows to other large Asian cities (e.g., Hong Kong, Singapore and Tokyo) grew dramatically in the 1990s.

Despite the reluctance to admit foreign labor, particularly low-skilled labor, a small but growing number of migrant workers have entered Seoul since the early 1990s. South Korea's economic success became common knowledge throughout Asia in the late 1980s, which made its capital city a new destination of inter-national labor migration. Recognizing the growing demand for cheap labor, the South Korean government has adopted a limited policy of bringing in guest workers who help relieve labor-shortage problems but are expected to return home upon the completion of their contracts. Currently, the government issues two kinds of labor certificate for migrant workers. The H-2 visa, under the Employment Management System, is issued to ethnic Korean migrants returning mostly from China. It allows these emigrants to work up to three years in the agricultural and construction sectors, as well as retailing and distribution service jobs. In the meantime, the Employment Permit System issues all other migrant workers – mostly Southeast and South Asian migrants – E-9 visas for three years of employ-ment in the manufacturing sector. Each year the government determines the number of work visas to be issued, based on the rule that the migrant labor force stays under 2 percent of the country's total workforce.

The conventional wisdom that strict (im)migration regulations at a time of substantial labor shortages in low-skilled jobs could encourage illegal immi-gration from labor-surplus countries has proven true in South Korea over the past decade. The number of illegal migrants who entered South Korea legally but have overstayed their temporary work or tourist visas has grown rapidly in recent years, now accounting for almost half of the country's 450,000 migrant workers (Figure 10.2).

The vast majority of migrant workers, both legal and illegal, work and reside in the metropolitan area of Seoul. While ethnic Koreans tend to settle in Seoul itself, where the demands for restaurant workers and daily laborers are high, other migrant workers gravitate to the peripheries of the metropolitan area, called Outer Seoul, where small and medium-sized manufacturing firms are concentrated.

Ethnic Korean migrants, most of whom are children and grandchildren of Korean peasants who migrated to northeast China during the Japanese colonial period (1910–45), have formed a migrant enclave in the Kuro area in the southwestern part of Seoul. The South Korean government is contemplating granting them Korean citizenship, but it will still take years for them to be recognized as part of their ethnic homeland. Since the bulk of Southeast and South Asian migrant workers are housed in dorm-like accommodation near their workplace, distinctive ethnic enclaves have not yet formed for any particular group. However, some

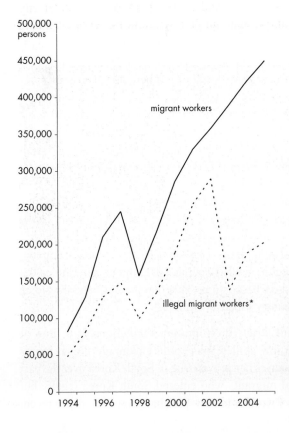

Figure 10.2
Migrant workforce in South Korea

Note: * Workers who have overstayed their visas

Source: Immigration Bureau, Republic of Korea (2006)

ethnic business districts have developed near the subway stations of industrial cities in Outer Seoul. Ansan, for example, home to around 5,000 manufacturing firms, has developed the country's largest Southeast Asian shopping street in Wonkok Dong, where a number of ethnic grocery stores and restaurants, bars, internet cafés and game clubs cluster together (Plate 10.2).

National and local officials and politicians across the political spectrum take pleasure in comparing Seoul to other large cities in terms of world city-ness. Much attention is paid to how they have hosted international sports competitions, upgraded infrastructure and increased foreign investment. Little is said about how many migrant workers they have received, how migrant workers have contributed to their cosmopolitan culture, or what policy provisions have been drafted to protect migrant workers from potentially exploitative situations. Instead, local media accounts associate the frequent sight of migrant workers in Outer Seoul with crime and prostitution. And national media repeatedly portrays the ethnic shopping district in Ansan as an exotic but high-crime area, which prompted local politicians to vow better policing for the benefit of "ordinary" residents.

Third world cities literature has long cited Seoul as a primate capital city preoccupied with over-urbanization issues, traffic congestion and many other third

Plate 10.2 Ethnic business center in Outer Seoul, South Korea

world problems. The city now appears regularly in the league tables of world cities, as many of its locally based multinational firms and banks rank in the Fortune Global 500. The South Korean government's world city projects, including hosting the Olympics and the World Cup, have certainly helped Seoul's graduation from third world city status to world city status. However, its world city-ness is highly skewed toward an economically globalizing world, and it is nowhere near cultural globalization. While newly forming migrant communities in Outer Seoul could make significant contributions to Seoul's world city-ness in terms of cosmopolitanism and multiculturalism, the South Korean government still looks to improve other aspects of Seoul's world city-ness. It remains to be seen whether the government will place cosmopolitanism, multiculturalism, openness, tolerance and social inclusion at the center of any future strategies for cementing Seoul's world city status.

Further reading

Davis, Mike, 2006, *Planet of Slums*, New York: Verso. A well-known author who has written several books on urban issues writes of the burgeoning growth of slums in third world cities and the forces that have worsened their poverty and inequality problems in recent years in a very accessible style.

Graham, Stephen and Simon, Marvin, 2001, *Splintering Urbanism: Networked Infrastructures, Technological Mobilities and the Urban Condition*, London: Routledge. This book examines how new transportation and telecommunications infrastructures have contributed to the splintering of metropolitan areas across the world.

United Nations Human Settlements Programme (UN-HABITAT), 2003, *The Challenge of Slums: Global Report on Human Settlements 2003*, London: Earthscan. This report is the first global assessment of urban slums, where almost 1 billion people currently live across the world. It presents not only challenging conditions in slums but policy responses and actions that could improve the livelihoods of slum-dwellers.

References

Abrahamson, Mark, 2004, *Global Cities*, New York: Oxford University Press.

Abu-Lughod, J. L., 1989, *Before European Hegemony: The World System, AD 1250–1350*, New York: Oxford University Press.

Aguiar, Luis L. M. and Andrew Herod, eds, 2007, *The Dirty Work of Neoliberalism: Cleaners in the Global Economy*, Malden: Blackwell.

Alderson, Arthur S. and Jason Beckfield, 2004, "Power and position in the world city system," *American Journal of Sociology*, 109(4), pp. 811–851.

Alter, Robert, 2005, *Imagined Cities: Urban Experience and the Language of the Novel*, New Haven: Yale University Press.

Amirahmadi, Hooshang and Salah S. El-Shakhs, eds, 1993, *Urban Development in the Muslim World*, New Brunswick: Center for Urban Policy Research.

Anderson, Benedict, 1983, *Imagined Communities: Reflections on the Origin and Spread of Nationalism*, London: Verso.

Appadurai, Arjun, 1996, *Modernity at Large: Cultural Dimensions of Globalization*, Minneapolis: University of Minnesota Press.

Ball, John Clement, 2004, *Imagining London: Postcolonial Fiction and the Transnational Metropolis*, Toronto: University of Toronto Press.

Barlow, Max and Brian Slack, 1985, "International cities: some geographical considerations and a case study of Montreal," *Geoforum*, 16(3), pp. 333–345.

Barter, James H., 2004, "Moscow's changing fortunes under three regimes," in Josef Gugler, ed., *World Cities beyond the West: Globalization, Development and Inequality*, Cambridge: Cambridge University Press, pp. 191–211.

Bascom, William, 1955, "Urbanization among the Yoruba," *American Journal of Sociology*, 60(5), pp. 446–454.

Beauregard, Robert A., 2006, *When America Became Suburban*, Minneapolis: University of Minnesota Press.

Beaverstock, J. V., R. G. Smith and P. J. Taylor, 1999, "A roster of world cities," *Cities*, 16(6), pp. 445–458.

Beaverstock, J. V., R. G. Smith, and P. J. Taylor, 2000, "Geographies of globalization: United States law firms in world cities," *Urban Geography*, 21, pp. 95–120.

Beavon, Keith S. O., 1997, "Johannesburg: a city and metropolitan area in transformation," in Carole Rakodi, ed., *The Urban Challenge in Africa: Growth and Management of Its Large Cities*, Tokyo: United Nations University Press, pp. 150–191.

Beavon, Keith S. O., 2006, "Johannesburg 1986–2030: a quest to regain world status," in M. Mark Amen, Kevin Archer and M. Martin Bosman, eds, *Relocating Global Cities: From the Center to the Margins*, Lanham: Rowman and Littlefield, pp. 49–73.

Bell, Daniel, 1973, *The Coming of Post-Industrial Society*, New York: Basic Books.

Benjamin, Solomon, 2000, "Governance, economic settings and poverty in Bangalore," *Environment and Urbanization*, 12(1), pp. 35–56.

Benton-Short, Lisa, Marie D. Price and Samantha Friedman, 2005, "Globalization from below: the ranking of global immigrant cities," *International Journal of Urban and Regional Research*, 29(4), pp. 945–959.

Benyon, H., 1973, *Working for Ford*, London: Allen Lane.

Berry, B. J. L., 1964, "Cities as systems within systems of cities," *Papers and Proceedings of the Regional Science Association*, 13, pp. 147–163.

Bishop, Ryan, John Phillips and Wei Wei Yeo, eds, 2003, *Postcolonial Urbanism: Southeast Asian Cities and Global Processes*, New York: Routledge.

Borchert, J. R., 1967, "American metropolitan evolution," *Geographical Review*, 57, pp. 301–332.

Boschma, R. A. and K. Frenken, 2006, "Why is economic geography not an evolutionary science? Towards an evolutionary economic geography," *Journal of Economic Geography*, 6, pp. 273–302.

Boyle, Mark, 2006, "Culture in the rise of tiger economies: Scottish expatriates in Dublin and the 'creative class' thesis," *International Journal of Urban and Regional Research*, 30(2), pp. 403–426.

Bradbury, Katharine L., Anthony Downs and Kenneth A. Small, 1982, *Urban Decline and the Future of American Cities*, Washington, DC: The Brookings Institution.

Brakman, Steven, Harry Garretsen and Charles van Marrewijk, 2001, *An Introduction to Geographical Economics: Trade, Location and Growth*, Cambridge: Cambridge University Press.

Braudel, Fernand, 1981–4, *Civilization and Capitalism 15th–18th Century*, London: William Collins' Sons.

Brenner, Neil and Roger Keil, eds, 2006, *The Global Cities Reader*, London: Routledge.

Brockerhoff, Martin and Ellen Brennan, 1998, "The poverty of cities in developing regions," *Population and Development Review*, 24(1), pp. 75–114.

Bromley, Ray, 1993, "Small-enterprise promotion as a urban development strategy," in John D. Kasarda and Allan M. Parnell, eds, *Third World Cities: Problems, Policies and Prospects*, Newbury Park: Sage, pp. 120–133.

Brown, D. A. and S. Ferino-Pagden, 2006, *Bellini, Giorgione, Titian and the Renaissance of Venetian Paintings*, New Haven: Yale University Press.

Bryceson, Deborah Fahy and Deborah Potts, eds, 2006, *African Urban Economies: Viability, Vitality or Vitiation?*, New York: Palgrave.

Buck, Nick, Ian Gordon, Alan Harding and Ivan Turok, eds, 2005, *Changing Cities: Rethinking Urban Competitiveness, Cohesion and Governance*, New York: Palgrave.

Budd, Leslie, 1999, "Globalization and the crisis of territorial embeddedness of international financial markets," in Ron Martin, ed., *Money and the Space Economy*, Chichester: John Wiley & Sons, pp. 115–137.

Bunnell, Tim, 2004, *Malaysia, Modernity and the Multimedia Super Corridor: A Critical Geography of Intelligent Landscapes*, London: Routledge.

Buttler, Friedrich (translated by John Cuthbert-Brown), 1975, *Growth Pole Theory and Economic Development*, Farnborough: Saxon House.

Cai, Jianming and Victor F. S. Sit, 2003, "Measuring world city formation – the case of Shanghai," *Annals of Regional Science*, 37(3), pp. 435–446.

Carter, Harold, 1983, *An Introduction to Urban Historical Geography*, London: Arnold.

Castells, Manuel, 2000, *The Rise of the Network Society* (second edition), Oxford: Blackwell.

Castells, Manuel and Alejandro Portes, 1989, "World underneath: the origins, dynamics, and effects of the informal economy," in Alejandro Portes, Manuel Castells and Lauren A. Benton, eds, *The Informal Economy: Studies in Advanced and Less Developed Countries*, Baltimore: Johns Hopkins University Press, pp. 11–37.

Chapman, Keith, 2005, "From 'growth centre' to 'cluster': restructuring, regional development, and the Teesside chemical industry," *Environment and Planning A*, 37(4), pp. 597–615.

Chase-Dunn, Christopher K., 1985, "The system of world cities: AD 800–1975," in Michael Timberlake, ed., *Urbanization in the World Economy*, New York: Academic Press, pp. 269–292.

Childe, V. Gordon, 1950, "The urban revolution," *Town Planning Review*, 21(1), pp. 3–17.

Christaller, W., 1966, *Central Places in Southern Germany*, Englewood Cliffs: Prentice-Hall.

Christerson, B. and C. Lever-Tracy, 1997, "The third China? Emerging industrial districts in rural China," *International Journal of Urban and Regional Research*, 21, pp. 569–584.

Clark, David, 1996, *Urban World/Global City*, London: Routledge.

Cohen, R. B., 1981, "The new international division of labour, multinational corporations and urban hierarchy," in Michael Dear and Allen J. Scott, eds, *Urbanization and Urban Planning in Capitalist Society*, New York: Methuen, pp. 287–315.

Connerly, C. E., 2005, *The Most Segregated City in America: City Planning and Civil Rights in Birmingham, 1928–1980*, Charlottesville: University of Virginia Press.

Cooke, P. and K. Morgan, 1998, *The Associational Economy*, Oxford: Oxford University Press.

Crane, Robert I., 1955, "Urbanism in India," *American Journal of Sociology*, 60(5), pp. 463–470.

Crosby, A. W., 1997, *The Measure of Reality: Quantification and Western Society*, Cambridge: Cambridge University Press.

Daniels, Peter, Andrew Leyshon, Mike Bradshaw and Jonathan Beaverstock, eds, 2007, *Geographies of the New Economy: Critical Reflections*, London: Routledge.

Davis, Diane E., 2005, "Cities in global context: a brief intellectual history," *International Journal of Urban and Regional Research*, 29(1), pp. 92–109.

Davis, Diane E. and Kian Tajbakhsh, 2005, "Globalization and cities in comparative perspective," *International Journal of Urban and Regional Research*, 29(1), pp. 89–91.

Davis, Mike, 1990, *City of Quartz*, New York: Vintage.

Davis, Mike, 2006, *Planet of Slums*, New York: Verso.

de Blij, Harm J., 1968, *Mombasa: An African City*, Chicago: Northwestern University Press.

Defoe, Daniel, 1928 [1728], *A Plan of the English Commerce*, Oxford: Blackwell.

De Soto, Hernando (translated by June Abbott), 1989, *The Other Path: The Invisible Revolution in the Third World*, New York: Harper & Row.

Dick, H. W. and P. J. Rimmer, 1998, "Beyond the third world city: the new urban geography of South-east Asia," *Urban Studies*, 35(12), pp. 2303–2321.

Dicken, Peter, 2007, *Global Shift: Mapping the Changing Contours of the World Economy*, New York: The Guilford Press.

Dollinger, P., 1970, *The German Hansa*, London: Macmillan.

Donck, A. van der, 1968 [1556], *A Description of New Netherlands*, Syracuse: Syracuse University Press.

Drakakis-Smith, David, 2000, *Third World Cities*, London: Routledge.

Dumenil, G. and D. Levy, 2004, (translated by D. Jeffers) *Capital Resurgent: Roots of the Neoliberal Revolution*, Cambridge, MA: Harvard University Press.

Dunning, John H., 2000, *Regions, Globalization, and the Knowledge-Based Economy*, Oxford: Oxford University Press.

Dziembowska-Kowalska, Jolanta and Rolf H. Funck, 1999, "Cultural activities: source of competitiveness and prosperity in urban regions," *Urban Studies*, 36(8), pp. 1381–1398.

Economist, The, 2000, "Cities and the Olympics: make or break," September 16, pp. 27–32.

Economist, The, 2006, "India's Muslims: don't blame it on the scriptures," December 2, p.48.

Economist, The, 2007, "The world goes to town: a special report on cities," May 5, pp. 1–18.

Ehrenreich, Barbara, 2001, *Nickel and Dimed: On (Not) Getting by in America*, New York: Metropolitan.

Engels, Friedrich, 1973 [1845], *The Condition of the Working Class in England*, Moscow: Progress.

Engels, Friedrich, 2004 [1848], *The Communist Manifesto*, Whitefish: Kessinger.

Estall, Robert C. and R. Ogilvie Buchanan, 1961, *Industrial Activity and Economic Geography*, London: Hutchinson.

Featherstone, Mike, 1995, *Undoing Culture: Globalization, Postmodernism and Identity*, Thousand Oaks: Sage.

Findley, Sally E., 1993, "The third world city: development policy and issues," in John D. Kasarda and Allan M. Parnell, eds, *Third World Cities: Problems, Policies and Prospects*, Newbury Park: Sage, pp. 1–31.

Florida, Richard, 2002, *The Rise of the Creative Class: And How It's Transforming Work, Leisure, Community and Everyday Life*, New York: Basic Books.

Florida, Richard, 2005a, *Cities and the Creative Class*, New York: Routledge.

Florida, Richard, 2005b, *The Flight of the Creative Class*, New York: Harper Business.

Foster, J., 1974, *Class Struggle and the Industrial Revolution*, London: Methuen.

Frank, Andre Gunder, 1966, "The development of underdevelopment," *Monthly Review*, September, pp. 17–30.

Friedmann, John, 1966, *Regional Development Policy: A Case Study of Venezuela*, Cambridge, MA: MIT Press.

Friedmann, John, 1986, "The world city hypothesis," *Development and Change*, 17, pp. 69–83.

Friedmann, John, 1995, "Where we stand: a decade of world city research," in Paul L. Knox and Peter J. Taylor, eds, *World Cities in a World System*, Cambridge: Cambridge University Press, pp. 21–47.

Friedmann, John, 2005, *China's Urban Transition*, Minneapolis: University of Minnesota Press.

Friedmann, John and Goetz Wolff, 1982, "World city formation: an agenda for research and action," *International Journal of Urban and Regional Research*, 6(3), pp. 309–344.

Fujita, Masahisa and Jacques-François Thisse, 2002, *Economics of Agglomeration: Cities, Industrial Location, and Regional Growth*, Cambridge, MA: Cambridge University Press.

Fujita, Masahisa, Paul Krugman, and Anthony J. Venables, 1999, *The Spatial Economy: Cities, Regions, and International Trade*, Cambridge: MIT Press.

Geddes, Patrick, 1915, *Cities in Evolution*, London: Williams and Norgate.

Gehrig, T., 1998, "Cities and the geography of financial centers," unpublished paper, Freiburg: University of Freiburg.

Gilbert, Alan, 1989, "Moving the capital of Argentina: a further example of utopian planning?," *Cities*, 6(3), pp. 234–242.

Ginsburg, Norton S., 1955, "The great city in Southeast Asia," *American Journal of Sociology*, 60(5), pp. 455–462.

Glaeser, Edward L., 2000, "The new economics of urban and regional growth," in Gordon L. Clark, Maryann P. Feldman and Meric S. Gertler, eds, *The Oxford Handbook of Economic Geography*, Oxford: Oxford University Press, pp. 83–98.

Glaeser, Edward L., Hedi D. Kallal, Jose A. Scheinkman and Andrei Shleifer, 1992, "Growth in cities," *Journal of Political Economy*, 100(6), pp. 1126–1152.

Glaeser, Edward L., Jose A. Scheinkman and Andrei Shleifer, 1995, "Economic growth in a cross-section of cities," *Journal of Monetary Economics*, 36(1), pp. 117–143.

Gospodini, A., 2002, "European cities in competition and the new 'uses' of urban design," *Journal of Urban Design*, 7(1), pp. 59–63.

Gottdiener, Mark and Chris G. Pickvance, eds, 1991, *Urban Life in Transition*, Newbury Park: Sage.

Gottmann, Jean and Robert A. Harper, eds, 1990, *Since Megalopolis: The Urban Writings of Jean Gottmann*, Baltimore: Johns Hopkins University Press.

Graham, Stephen and Simon Marvin, 2001, *Splintering Urbanism: Networked Infrastructures, Technological Mobilities and the Urban Condition*, London: Routledge.

Granovetter, Mark, 1985, "Economic action and social structure: the problem of embeddedness," *American Journal of Sociology*, 91, pp. 481–510.

Grant, Richard, 2001, "Liberalization policies and foreign companies in Accra, Ghana," *Environment and Planning A*, 33(6), pp. 997–1014.

Grant, Richard, 2002, "Foreign companies and glocalizations: evidence from Accra, Ghana," in Richard Grant and John Rennie Short, eds, *Globalization and the Margins*, New York: Palgrave, pp. 130–149.

Grant, Richard and Jan Nijman, 2002, "Globalization and the corporate geography of cities in the less-developed world," *Annals of the Association of American Geographers*, 92(2), pp. 320–340.

Grant, Richard and John Rennie Short, eds, 2002, *Globalization and the Margins*, New York: Palgrave.

Gugler, Josef, ed., 1988, *The Urbanization of the Third World*, Oxford: Oxford University Press.

Gugler, Josef, ed., 1996, *The Urban Transformation of the Developing World*, Oxford: Oxford University Press.

Gugler, Josef, ed., 1997, *Cities in the Developing World: Issues, Theory and Policy*, Oxford: Oxford University Press.

Gugler, Josef, ed., 2004, *World Cities beyond the West: Globalization, Development and Inequality*, Cambridge: Cambridge University Press.

Halfani, Mohamed, 1996, "Marginality and dynamism: prospects for the sub-Saharan African City," in Michael A. Cohen, Blair A. Ruble, Joseph S. Tulchin and Allison M. Garland, eds, *Preparing for the Urban Future: Global Pressures and Local Forces*, Washington, DC: Woodrow Wilson Center Press, pp. 83–107.

Hall, Peter, 1966, *The World Cities*, New York: World University Library.

Hall, Peter, 1988, *Cities of Tomorrow*, Oxford: Basil Blackwell.

Hall, Peter, 1993, "The changing role of capital cities: six types of capital city," in John Taylor, Jean G. Lengellé and Caroline Andrew, eds, *Capital Cities: International Perspectives*, Ottawa: Carleton University Press, pp. 69–84.

Hall, Peter, 1998, *Cities in Civilization*, London: Weidenfeld & Nicolson.

Hall, Peter and Kathy Pain, eds, 2006, *The Polycentric Metropolis: Learning from Mega-City Regions in Europe*, London: Earthscan.

Hall, Peter and Paschal Preston, 1988, *The Carrier Wave: New Information Technology and the Geography of Innovation, 1846–2003*, London: Unwin Hyman.

Hall, Tim and Phil Hubbard, eds, 1998, *The Entrepreneurial City: Geographies of Politics, Regime and Representation*, Chichester: John Wiley & Sons.

Hamm, Steve, 2007, *Bangalore Tiger: How Indian Tech Upstart Wipro Is Rewriting the Rules of Global Competition*, New York: McGraw-Hill.

Hancock, David, 1995, *Citizens of the World: London Merchants and the Integration of the British Atlantic Community, 1735–1785*, Cambridge: Cambridge University Press.

Hannerz, Ulf, 1996, *Transnational Connections: Culture, People, Places*, London: Routledge.

Hardoy, Jorge E., 1993, "Ancient capital cities and new capital cities of Latin America," in John Taylor, Jean G. Lengellé and Caroline Andrew, eds, *Capital Cities: International Perspectives*, Ottawa: Carleton University Press, pp. 99–128.

Harris, Nigel, 1978, *Economic Development, Cities, and Planning: The Case of Bombay*, New York: Oxford University Press.

Hartley, John, ed., 2005, *Creative Industries*, Oxford: Blackwell.

Harvey, David, 1989a, "From managerialism to entrepreneurialism: the transformation in urban governance in late capitalism," *Geografiska Annaler B*, 71, pp. 3–17.

Harvey, David, 1989b, *The Condition of Postmodernity: An Enquiry into the Origins of Cultural Change*, Cambridge: Blackwell.

Harvey, David, 2005, *A Brief History of Neoliberalism*, New York: Oxford University Press.

Hauser, Philip M., 1955, "World urbanism," *American Journal of Sociology*, 60(5), pp. 427–428.

Henderson, J. Vernon, 1988, *Urban Development: Theory, Fact, and Illusion*, Oxford: Oxford University Press.

Hoffman, Lily M., 2003, "The marketing of diversity in the inner city: tourism and regulation in Harlem," *International Journal of Urban and Regional Research*, 27(2), pp. 286–299.

Holston, James, 1989, *The Modernist City: An Anthropological Critique of Brasilia*, Chicago: The University of Chicago Press.

Hoover, E. M., 1948, *The Location of Economic Activity*, New York: McGraw-Hill.

Hurst, E., 1972, *A Geography of Economic Behavior*, Belmont: Wadsworth.

Jacobs, Jane, 1969, *The Economy of Cities*, New York: Random House.

Jameson, Fredric, 1991, *Postmodernism, or, the Cultural Logic of Late Capitalism*, Durham, NC: Duke University Press.

Jardine, Lisa, 1996, *Worldly Goods: A New History of the Renaissance*, London: Macmillan.

Jonas, Andrew and David Wilson, eds, 1999, *The Urban Growth Machine: Critical Perspectives Two Decades Later*, Albany: State University of New York Press.

Jordan, Jennifer, 2003, "Collective memory and locality in global cities," in Linda Krause and Patrice Petro, eds, *Global Cities: Cinema, Architecture, and Urbanism in a Digital Age*, New Brunswick: Rutgers University Press, pp. 31–48.

Joyce, Patrick, 1980, *Work, Society and Politics: The Culture of the Factory in Later Victorian England*, Brighton: Harvester Press.

Judd, Dennis and Paul Kantor, eds, 1992, *Enduring Tensions in Urban Politics*, New York: Macmillan.

Kasarda, John D. and Allan M. Parnell, eds, 1993, *Third World Cities: Problems, Policies and Prospects*, Newbury Park: Sage.

Katznelson, I., 1979, *City Trenches: Urban Politics and The Patterning of Class in the United States*, New York: Pantheon.

Keeling, David J., 1995, "Transport and the world city paradigm," in Paul L. Knox and Peter J. Taylor, eds, *World Cities in a World System*, Cambridge: Cambridge University Press, pp. 115–131.

Keeling, David J., 1996, *Buenos Aires: Global Dreams, Local Crises*, Chichester: John Wiley & Sons.

Kelley, Allen C. and Jeffrey G. Williamson, 1984, *What Drives Third World City Growth?: A Dynamic General Equilibrium Approach*, Princeton: Princeton University Press.

Keniston, Kenneth and Deepak Kumar, eds, 2004, *IT Experience in India: Bridging the Digital Divide*, New Delhi: Sage.

Kennedy, Paul, 1987, *The Rise and Fall of the Great Powers: Economic Change and Military Conflict from 1500 to 2000*, New York: Random House.

Keyder, Caglar, 2005, "Globalization and social exclusion in Istanbul," *International Journal of Urban and Regional Research*, 29(1), pp. 124–134.

Kim, J., S. Rhee, J. Yu, K. Ku and J. Hong, 1989, *The Impact of the Seoul Olympic Games on National Development*, Seoul: Korea Development Institute (in Korean).

Kim, Joochul and Sang-Chuel Choe, 1997, *Seoul: The Making of a Metropolis*, Chichester: John Wiley & Sons.

Kim, Yeong-Hyun, 1997, "Interpreting the Olympic landscape in Seoul: the politics of sports, spectacle and landscape," *Journal of the Korean Geographical Society*, 32(3), pp. 387–402.

Kim, Yeong-Hyun, 1998, *Globalization, Urban Changes and Seoul's Dreams – A Global Perspective on Contemporary Seoul*, Ph.D. dissertation, Department of Geography, Syracuse University, Syracuse, New York.

Kim, Yeong-Hyun, 2004, "Seoul: complementing economic success with Games," in Josef Gugler, ed., *World Cities beyond the West: Globalization, Development and Inequality*, Cambridge: Cambridge University Press, pp. 59–81.

King, Anthony D., 2004, *Spaces of Global Cultures: Architecture Urbanism Identity*, London: Routledge.

King, Kenneth, 2001, "Africa's informal economies: thirty years on," *SAIS Review of International Affairs*, 21(1), pp. 97–108.

Knox, Paul L., ed., 1993, *The Restless Urban Landscape*, Englewood Cliffs: Prentice Hall.

Knox, Paul L. and Linda McCarthy, 2005, *Urbanization: An Introduction to Urban Geography* (second edition), Upper Saddle River: Pearson Prentice Hall.

Knox, Paul L. and Peter J. Taylor, eds, 1995, *World Cities in a World System*, Cambridge: Cambridge University Press.

Kratke, Stefan, 2004, "City of talents?: Berlin's regional economy, socio-spatial fabric and 'worst practice' urban governance," *International Journal of Urban and Regional Research*, 28(3), pp. 511–529.

Krause, Linda and Patrice Petro, eds, 2003, *Global Cities: Cinema, Architecture, and Urbanism in a Digital Age*, New Brunswick: Rutgers University Press.

Kresl, Peter Karl and Earl H. Fry, 2005, *The Urban Response to Internationalization*, Cheltenham: Edward Elgar.

Landry, Charles, 2000, *The Creative City: A Toolkit for Urban Innovators*, London: Earthscan.

Laquian, Aprodicio A., 2005, *Beyond Metropolis: The Planning and Governance of Asia's Mega-Urban Regions*, Washington, DC: Woodrow Wilson Center Press.

Larson, James F. and Heung-Soo Park, 1993, *Global Television and the Politics of the Seoul Olympics*, Boulder: Westview Press.

Lash, Scott and John Urry, 1994, *Economies of Signs and Space*, London: Sage.

Lasuen, J. R., 1969, "On growth poles," *Urban Studies*, 6(2), pp. 137–161.

Lauria, Mickey, 1997, *Reconstructing Urban Regime Theory: Regulating Urban Politics in a Global Economy*, London: Sage.

Le, L., 2004, "Is Shanghai really a 'global city'?," unpublished paper presented in the City Futures Conference, Chicago, July 8–10.

Linsky, Arnold, 1965, "Some generalizations concerning primate cities," *Annals of the Association of American Geographers*, 55(3), pp. 506–513.

Lo, Fu-chen and Yue-man Yeung, eds, 1996, *Emerging World Cities in Pacific Asia*, Tokyo: United Nations University Press.

Logan, John R., ed., 2002, *The New Chinese City: Globalization and Market Reform*, Oxford: Blackwell.

Logan, John R. and Harvey L. Molotch, 1987, *Urban Fortunes: The Political Economy of Place*, Berkeley: University of California Press.

Lowder, Stella, 1986, *The Geography of Third World Cities*, Totowa: Barnes & Noble Books.

Lucas, Glenn C., 1988, "On the mechanics of economic development," *Journal of Monetary Economics*, 22, pp. 3–42.

McCann, Eugene J., 2004, "Urban political economy beyond the 'global city'," *Urban Studies*, 41(12), pp. 2315–2333.

McGee, T.G., 1967, *The Southeast Asian City: A Social Geography of the Primate Cities of Southeast Asia*, New York: Frederick A. Praeger.

McGeehan, Patrick, 2006, "Top executives return offices to Manhattan," *New York Times*, July 3rd (<http://nytimes.com/>).

Machimura, Takashi, 1998, "Symbolic use of globalization in urban politics in Tokyo," *International Journal of Urban Regional Research*, 22(2), pp. 183–194.

Madanipour, Ali, 1998, *Tehran: The Making of a Metropolis*, Chichester: John Wiley & Sons.

Markusen, Ann, 2006, "Urban development and the politics of a creative class: evidence from a study of artists," *Environment and Planning A*, 38(10), pp. 1921–1940.

Markusen, Ann and Greg Schrock, 2006, "The artistic dividend: urban artistic specialization and economic development implications," *Urban Studies*, 43(10), pp. 1661–1686.

Martin, R. L., 2000, "Institutional approaches to economic geography," in D. Sheppard and T. J. Barnes, eds, *A Companion to Economic Geography*, Oxford: Blackwell, pp. 77–94.

Martin, Ron, 1999, *Money and the Space Economy*, Chichester: John Wiley & Sons.

Matson, Cathy, 1997, *Merchants and Empire: Trading in Colonial New York*, Baltimore: Johns Hopkins University Press.

Mehta, Suketu, 2004, *Maximum City: Bombay Lost and Found*, New York: Knopf.

Miao, Pu, 2003, "Deserted streets in a jammed town: the gated community in Chinese cities and its solution," *Journal of Urban Design*, 8(1), pp. 45–66.

Moreau, Ron and Sudip Mazumdar, 2004, "A change of address: the outsourcing industry is moving to cheaper locales," *Newsweek International Edition*, September 27th (<http://www.msnbe.msn.com/id/6038643/site/newsweek/>).

Mossberger, Karen and Gerry Stoker, 2001, "The evolution of urban regime theory: the challenge of conceptualization," *Urban Affairs Review* 36(6), pp. 810–835.

Mumford, Lewis, 1961, *The City in History: Its Origins, Its Transformations, and Its Prospects*, New York: MJF Books.

Murray, Martin J., 2004, *The Evolving Spatial Form of Cities in a Globalising World Economy: Johannesburg and São Paulo*, Democracy and Governance Research Programme, Occasional Paper 5, Cape Town: Human Sciences Research Council.

Murray, Warwick E., 2006, *Geographies of Globalization*, London: Routledge.

Mycoo, Michelle, 2006, "The retreat of the upper and middle classes to gated communities in the poststructural adjustment era: the case of Trinidad," *Environment and Planning A*, 38(1), pp. 131–148.

Naisbitt, John, 1994, *Global Paradox: The Bigger the World Economy, the More Powerful Its Smallest Players*, New York: W. Morrow.

Negroponte, Nicholas, 1995, *Being Digital*, New York: Knopf.

Neuwirth, R., 2004, *Shadow Cities: A Billion Squatters, a New Urban World*, London: Routledge.

Nichols, T. and Benyon, H., 1977, *Living with Capitalism: Class Relations and the Modern Factory*, London: Routledge.

O'Flaherty, Brendan, 2005, *City Economics*, Cambridge, MA: Harvard University Press.

Oberai, A. S., 1993, *Population Growth, Employment and Poverty in Third-World Mega-Cities: Analytical and Policy Issues*, New York: St. Martin's Press.

Öncü, Ayşe and Peter Weyland, eds, 1997, *Space, Culture and Power: New Identities in Globalizing Cities*, London: Zed Books.

Oren, Tasha G., 2003, "Gobbled up and gone – cultural preservation and the global city market place," in Linda Krause and Patrice Petro, eds, *Global Cities: Cinema, Architecture, and Urbanism in a Digital Age*, New Brunswick: Rutgers University Press, pp. 49–68.

Pacione, Michael, 2001, *Urban Geography: A Global Perspective*, London: Routledge.

Parker, G., 2006, "The megacity: decoding the chaos of Lagos," *New Yorker*, November 13, pp. 62–75.

Parnell, Susan, 2007, "Politics of transformation: defining the city strategy in Johannesburg," in Klaus Segbers, ed., *The Making of Global City Regions: Johannesburg, Mumbai/Bombay, São Paulo, and Shanghai*, Baltimore: Johns Hopkins University Press, pp. 139–167.

Parr, John B., 1999, "Growth-pole strategies in regional economic planning: a retrospective view. Part 1: origins and advocacy," *Urban Studies*, 36(7), pp. 1195–1215.

Parthasarathy, Balaji, 2004, "India's Silicon Valley or Silicon Valley's India? Socially embedding the computer software industry in Bangalore," *International Journal of Urban and Regional Research*, 28(3), pp. 664–685.

Paul, Darel E., 2004, "World cities as hegemonic projects: the politics of global imagineering in Montreal," *Political Geography*, 23(5), pp. 571–596.

Peck, Jamie, 2005, "Struggling with the creative class," *International Journal of Urban Regional Research*, 29(4), pp. 740–770.

Peters, Alan and Peter Fisher, 2004, "The failures of economic development incentives," *Journal of the American Planning Association*, 70(1), pp. 27–37.

Piven, F. F. and Cloward R. A., 1997, *The Breaking of the American Social Compact*, New York: Free Press.

Portes, Alejandro, Manuel Castells and Lauren A. Benton, eds, 1989, *The Informal Economy: Studies in Advanced and Less Developed Countries*, Baltimore: Johns Hopkins University Press.

Potter, Robert B., 1992, *Urbanization in the Third World*, Oxford: Oxford University Press.

Potter, Robert B., Tony Binns, Jennifer A. Elliott, and David Smith, 1999, *Geographies of Development*, Harlow: Longman.

Potter, Robert B. and Sally Lloyd-Evans, 1998, *The City in the Developing World*, Harlow: Longman.

Powell, John A., 2004, "Civil rights, sprawl and regional equity," paper presented in the Columbus Metropolitan Club Forum (<www.kirwaninstitute.org/multimedia/presentations/2004_06_09_ColMetroclub.ppt>).

Power, Dominic and Allen J. Scott, eds, 2004, *Cultural Industries and the Production of Culture*, London: Routledge.

Pred, A. R., 1966, *The Spatial Dynamics of US Urban–Industrial Growth 1800–1914*, Cambridge, MA: MIT Press.

Quilley, Stephen, 2000, "Manchester first: from municipal socialism to the entrepreneurial city," *International Journal of Urban and Regional Research*, 24(3), pp. 601–615.

Rai, Saritha, 2006, "Is the next Silicon Valley taking root in Bangalore?," *New York Times*, March 20 (<http://www.nytimes.com>).

Rakodi, Carole, ed., 1997, *The Urban Challenge in Africa: Growth and Management of Its Large Cities*, Tokyo: United Nations University Press.

Rantisi, Norma M., 2004, "The ascendance of New York fashion," *International Journal of Urban and Regional Research*, 28(1), pp. 86–106.

Reich, Robert B., 1991, *The Work of Nations*, New York: Knopf.

Richardson, Harry W., 1981, "National urban development strategies in developing countries," *Urban Studies*, 18(3), pp. 267–283.

Richardson, Harry W., 1987, "Whither national urban policy in developing countries?," *Urban Studies*, 24(3), pp. 227–244.

Roberts, Bryan R., 1978, *Cities of Peasants: The Political Economy of Urbanization in the Third World*, Beverly Hills: Sage.

Roberts, Bryan R., 2005, "Globalization and Latin American cities," *International Journal of Urban Regional Research*, 29(1), pp. 110–123.

Robinson, Jennifer, 2002, "Global and world cities: a view from off the map," *International Journal of Urban and Regional Research*, 26(3), pp. 531–554.

Robinson, Jennifer, 2006, *Ordinary Cities: Between Modernity and Development*, New York: Routledge.

Romer, Paul M., 1986, "Increasing returns and long-run growth," *Journal of Political Economy*, 94, pp. 1002–1037.

Rondinelli, Dennis A., 1983, *Secondary Cities in Developing Countries: Policies for Diffusing Urbanization*, Beverly Hills: Sage.

Rosenberg, David, 2002, *Cloning Silicon Valley: The Next Generation High-tech Hotspots*, London: Reuters.

Sachs, Jeffrey D., 2005, *The End of Poverty: Economic Possibilities for Our Time*, New York: Penguin Press.

Sasaki, Masayuki, 2004, "The role of culture in urban regeneration," paper presented in Universal Forum of Cultures, Barcelona (<www.barcelona2004.org/esp/banco_del_ conocimiento/docs/PO_22_EN_SASAKI.pdf>).

Sassen, Saskia, 1991, *The Global City: New York, London, Tokyo*, Princeton: Princeton University Press.

Sassen, Saskia, 1994, *Cities in a World Economy*, Thousand Oaks: Forge Press.

Sassen, Saskia, ed., 2002, *Global Networks, Linked Cities*, New York: Routledge.

Savitch, H. V., 1996, "Cities in a global era: a new paradigm for the next millennium," in Michael A. Cohen, Blair A. Ruble, Joseph S. Tulchin and Allison M. Garland, eds, *Preparing for the Urban Future: Global Pressures and Local Forces*, Washington, DC: Woodrow Wilson Center Press, pp. 39–65.

Savitch, H. V. and Paul Kantor, 2002, *Cities in the International Marketplace: The Political Economy of Urban Development in North America and Western Europe*, Princeton: Princeton University Press.

Saxenian, AnnaLee, 2004, "The Bangalore boom: from brain drain to brain circulation", in Kenneth Keniston and Deepak Kumar, eds, *IT Experience in India: Bridging the Digital Divide*, New Delhi: Sage, pp. 169–181.

Saxenian, AnnaLee, 2006, *The New Argonauts: Regional Advantage in a Global Economy*, Cambridge, MA: Harvard University Press.

Schama, S., 1987, *The Embarrassment of Riches*, London: Collins.

Scott, Allen J., 2000, *The Cultural Economy of Cities: Essays on the Geography of Image-Producing Industries*, London: Sage.

Scott, Allen J., 2005, *On Hollywood: The Place, the Industry*, Princeton: Princeton University Press.

Scott, Allen J., 2006, *Geography and Economy: Three Lectures*, Oxford: Oxford University Press.

Seabrook, Jeremy, 1996, *In the Cities of the South: Scenes from a Developing World*, London: Verso.

Segbers, Klaus, ed., 2007, *The Making of Global City Regions: Johannesburg, Mumbai/Bombay, São Paulo, and Shanghai*, Baltimore: Johns Hopkins University Press.

Segre, Roberto, Mario Goyula and Joseph L. Scarpaci, 1997, *Havana: Two Faces of the Antillean Metropolis*, Chichester: John Wiley & Sons.

Sethuraman, S. V., 1997, "Urban poverty and the informal sector: a critical assessment of current strategies," International Labor Organization (<http://www-ilo-mirror. cornell.edu/public/english/employment/recon/eiip/publ/1998/urbpover.htm>).

Shenoy, Bhamy V., 2006, "Beware the Dutch disease, Bangalore," *Vijay Times Bangalore*, April 2.

Short, John Rennie, 1996, *The Urban Order*, Oxford: Blackwell.

Short, John Rennie, 2001, *Representing the Republic: Mapping the US, 1600–1900*, London: Reaktion.

Short, John Rennie, 2004a, *Global Metropolitan: Globalizing Cities in a Capitalist World*, London: Routledge.

Short, John Rennie, 2004b, *Making Space: Revisioning the World, 1475–1600*, Syracuse: Syracuse University Press.

Short, John Rennie, 2006a, *Alabaster Cities*, Syracuse: Syracuse University Press.

Short, John Rennie, 2006b, *Urban Theory: A Critical Assessment*, New York: Palgrave.

Short, J. R., Y. Kim, M. Kuus and H. Wells, 1996, "The dirty little secret of world cities research: data problems in comparative analysis," *International Journal of Urban and Regional Research*, 20(4), pp. 697–717.

Short, John Rennie and Yeong-Hyun Kim, 1999, *Globalization and the City*, New York: Longman.

Siemiatycki, Matti, 2006, "Message in a metro: building urban rail infrastructure and image in Delhi, India," *International Journal of Urban and Regional Research*, 30(2), pp. 277–292.

Simone, AbdouMaliq, 2004, *For the City yet to Come: Changing African Life in Four Cities*, Durham, NC: Duke University Press.

Sklair, Leslie, 2005, "The transnational capitalist class and contemporary architecture in glabalizing cities," *International Journal of Urban and Regional Research*, 29(3), pp. 485–500.

Smith, David A., 1996, *Third World Cities in Global Perspective: The Political Economy of Uneven Urbanization*, Boulder: Westview Press.

Smith, David A., 2003, "Rediscovering cities and urbanization in the 21st century world-system," in Wilma A. Dunaway, ed., *Emerging Issues in the 21st Century World-System, Volume II: New Theoretical Directions for the 21st Century World-System*, Westport: Praeger, pp. 111–129.

Smith, David and Michael Timberlake, 1995, "Cities in global matrices: toward mapping the world-system's city system," in Paul L. Knox and Peter J. Taylor, eds, *World Cities in a World-System*, Cambridge: Cambridge University Press, pp. 79–97.

Smith, David M., ed., 1992, *The Apartheid City and beyond: Urbanization and Social Change in South Africa*, London: Routledge.

Smith, Michael P., 2001, *Transnational Urbanism: Locating Globalization*, Oxford: Blackwell.

Smith, Michael Peter and Joe R. Feagin, eds, 1987, *The Capitalist City: Global Restructuring and Community Politics*, Oxford: Blackwell.

Smith, Neil, 1996, *The New Urban Frontier: Gentrification and the Revanchist City*, London: Routledge.

Southey, R., 1951 [1807], *Letters from England by Don Manuel Alvarez Espriella*, Vol. I, London: Cresset.

Stoker, Gerry and Karen Mossberger, 1994, "Urban regime theory in comparative perspective," *Environment and Planning C*, 12(2), pp. 195–212.

Stone, Clarence N., 1989, *Regime Politics: Governing Atlanta 1946–1988*, Lawrence: University Press of Kansas.

Stone, Clarence N., 2005, "Looking back to look forward; reflections on urban regime analysis," *Urban Affairs Review*, 40(3), pp. 309–341.

Storper, M., 1997, *The Regional World*, New York: Guildford.

Storper, M. and Venables, A. J., 2004, "Buzz: face to face contact and the urban economy," *Journal of Economic Geography*, 4, pp. 351–370.

Sutcliffe, Anthony, 1993, "Capital cities: does form follow values?," in John Taylor, Jean G. Lengellé and Caroline Andrew, eds, *Capital Cities: International Perspectives*, Ottawa: Carleton University Press, pp. 195–212.

Taaffe, E. J., R. L. Morrill and P. R. Gould, 1963, "Transport expansion in underdeveloped countries: a comparative analysis," *Geographical Review*, 53(3), pp. 503–529.

Tao, Z. and Y. C. R. Wong, 2002, "Hong Kong: from an industrialized city to a centre of manufacturing-related services," *Urban Studies*, 39(12), pp. 2345–2358.

Tarver, James D., 1994, *Urbanization in Africa: A Handbook*, Westport: Greenwood Press.

Taylor, John, Jean G. Lengellé and Caroline Andrew, eds, 1993, *Capital Cities: International Perspectives*, Ottawa: Carleton University Press.

Taylor, Peter J., 2000, "World cities and territorial states under conditions of contemporary globalization," *Political Geography*, 19(1), pp. 5–32.

Taylor, Peter J., 2004, *World City Network: A Global Urban Analysis*, London: Routledge.

Taylor, Peter J., 2005, "Leading world cities: empirical evaluations of urban nodes in multiple networks," *Urban Studies*, 42(9), pp. 1593–1608.

Taylor, Peter J., D. R. F. Walker and J. V. Beaverstock, 2002, "Firms and their global service networks," in Saskia Sassen, ed., *Global Networks, Linked Cities*, New York: Routledge, pp. 93–115.

Thompson, E. P., 1963, *The Making of the English Working Class*, Harmondsworth: Penguin.

Thompson, W., 1965, *Preface to Urban Economics*, Baltimore: Johns Hopkins University Press.

Thrift, Nigel, 1994, "On the social and cultural determinants of international financial centres: the case of the City of London," in Stuart Corbridge, Ron Martin and Nigel Thrift, eds, *Money, Power and Space*, Oxford: Blackwell, pp. 327–355.

Throsby, David, 2001, *Economics and Culture*, Cambridge: Cambridge University Press.

Timberlake, Michael, ed., 1985, *Urbanization in the World-Economy*, New York: Academic Press.

Todaro, Michael P. and Stephen C. Smith, 2003, *Economic Development*, London: Addison Wesley (Chapter 8: "Urbanization and rural-migration: theory and policy").

United Nations, 2005, *Population Challenges and Development Goals*, New York: United Nations.

United Nations Conference on Trade and Development (UNCTAD), 2005, *The Digital Divide: ICT Development Indices 2004*, New York: United Nations (<http://stdev.unctad.org/docs/digitaldivide.doc>).

United Nations Human Settlements Programme (UN-HABITAT), 2003, *The Challenge of Slums: Global Report on Human Settlements 2003*, London: Earthscan.

United Nations Human Settlements Programme (UN-HABITAT), 2006–7, *The State of the World's Cities Report 2006/2007: The Millennium Development Goals and Urban Sustainability*, London: Earthscan.

United Nations, Population Division, 1995, *The Challenge of Urbanization: The World's Largest Cities*, New York: United Nations.

Ward, Peter M., 1998, *Mexico City*, Chichester: John Wiley & Sons.

Webber, Michael J. and David L. Rigby, 1996, *The Golden Age Illusion: Rethinking Postwar Capitalism*, New York: Guilford.

Webster, Chris, Georg Glasze and Klaus Frantz, 2002, "The global spread of gated communities," *Environment and Planning B*, 29(3), pp. 315–320.

Werna, Edmundo, 2000, *Combating Urban Inequalities: Challenges for Managing Cities in the Developing World*, Cheltenham: Edward Elgar.

Williams, Raymond, 1973, *The Country and the City*, London: Cox & Wyman.

Wilson, William Julius, 1996, *When Work Disappears: The World of the New Urban Poor*, New York: Knopf.

Wolman, Harold, Edward W. Hill and Kimberly Furdell, 2004, "Evaluating the success of urban success stories: is reputation a guide to best practice?," *Housing Policy Debate*, 15(4), pp. 965–997.

Wong, J. and S. Chan, 2002, "China's emergence as a global manufacturing centre: implications for ASEAN," *Asia Pacific Business Review*, 9, pp. 79–94.

Wrigley, E. A., 1967, "A simple model of London's importance in changing English society and economy 1650–1750," *Past and Present*, 37, pp. 44–70.

Wu, Fulong, 2003, "Transition cities," *Environment and Planning A*, 35(8), pp. 1331–1338.

Xu, Jiang and Anthony G. O. Yeh, 2005, "City repositioning and competitiveness building in regional development: new development strategies in Guangzhou, China," *International Journal of Urban and Regional Research*, 29(2), pp. 283–308.

Yap, Jen Yih, 2004, *From a Capital City to a World City: Vision 2020, Multimedia Super Corridor and Kuala Lumpur*, masters thesis, Center for International Studies, Ohio University.

Yeoh, Brenda S. A., 1999, "Global/globalizing cities," *Progress in Human Geography*, 23(4), pp. 607–616.

Yeoh, Brenda S. A., 2001, "Postcolonial cities," *Progress in Human Geography*, 25(3), pp. 456–468.

Yeoh, Brenda S. A. and T. C. Chang, 2001, "Globalising Singapore: debating trans-national flows in the city," *Urban Studies*, 38(7), pp. 1025–1044.

Yusuf, Shahid and Kaoru Nabeshima, 2006, *Postindustrial East Asian Cities: Innovation for Growth*, Washington, DC: World Bank.

Zhao, S. X., 2003, "Spatial restructuring of financial centers in mainland China and Hong Kong: a geography of finance perspective," *Urban Affairs Review*, 38(4), pp. 535–571.

Zoványi, G, 1989, "The evolution of a national urban development strategy in Hungary" *Environment and Planning A*, 21(3), pp. 333–347.

Zukin, Sharon, 1995, *The Cultures of Cities*, Cambridge: Blackwell.

Data sources

Airports Council International, 2006, "The Voice of the world's airports" (<http://www.airports.org>).

Akron, City of, 2006, "Mayor's Office of Economic Development" (<http://www.ci.akron.oh.us/ed/index.htm>).

Atlanta, City of, 2005, "New century economic development plan for the City of Atlanta" (<http://www.atlantada.com/media/EDPRevisionAugust05.pdf>).

BBC, 2007, "Bangalore's boomtown blues," January 29 (<http:news.bbc.co.uk>).

City Mayors, 2006, "The largest US cities" (<http://www.citymayors.com/gratis/uscities_100.html>).

CNNMoney.com, 2006, "Fortune Global 500" (<http://money.cnn.com/magazines/fortune/global500/2006/cities/>).

CreativeTampaBay.com, 2007, "History" (<http://www.creativetampabay.com/>).

Denver, City of, 2006, "The Denver Office of Economic Development" (<http://www.milehigh.com/>).

Emporis Buildings, 2006, "Official world's 200 tallest high-rise buildings" (<http://www.emporis.com/en/bu/>).

Glasgow, City of, 2006, "Our vision for Glasgow" (<http://www.glasgow.gov.uk/>).

Immigration Bureau, Ministry of Justice, Republic of Korea, 2006, "Major policy guide" (<http://www.immigration.go.kr/>).

International Olympic Committee, 2006, "All the Games since 1896" (<http://www.olympic.org>).

Internet Modern History Sourcebook, The, 1997, "Industrial revolution" (<http://www.fordham.edu/halsall.mod/modsbook14.html>).

Johannesburg, City of, 2007, "Joburg 2030" (<http://www.joburg.org.za/>).

Joint Venture: Silicon Valley Network, 2007, *2007 Index of Silicon Valley* (<http://www.jointventure.org/publicatons/publicatons.html>).

Mercer Human Resource Consulting, 2007, "Highlights from the *Quality of Living Survey*" (<http://www.mercerhr.com>).

National Statistical Office, Republic of Korea, 2006, "2005 Census" (<http://kosis.nso.go.kr>).

Osaka, City of, 2006, "Creative city initiative" (<http://www.city.osaka.jp/english/mayors_message/conference/2006_04_26.html>).

Overseas Korean Foundation, 2006, "Statistics of Overseas Koreans" (<http://www.okf.or.kr/data/abodeStatus.jsp>).

Program on Globalization and Regional Innovation System, University of Toronto,

2006, "Imagine a Toronto . . . strategies for a creative city" (<http://www.imagine toronto.ca>).

San Francisco, City of, 2006, "Mayor's Office of Economic and Workforce Development" (<http://www.sfgov.org/site/moed_index.asp>).

Seoul Metropolitan Government, 2006, "International Cooperation: Policy Directions and Goals" (<http://english.seoul.go.kr/gover/cooper/coo_01pol.html>).

Toledo, City of, 2006, "Our mission" (<http://www.ci.toledo.oh.us>).

United Nations, 2006, "Population density and urbanization: population of capital cities and cities of 100,000 and more inhabitants" (<http://unstats.un.org/unsd/demographic/sconcerns/densurb/urban.aspx>).

United Nations Human Settlements Programme (UN-HABITAT), 2007, "Global urban observatory databases" (<http://www.devinfo.info/urbaninfo/>).

US Census Bureau (Campbell Gibson), 1998, "Population of the 100 largest cities and other urban places in the United States: 1790 to 1990" (<http://www.census.gov/population/www/documentation/twps0027.html>).

US Census Bureau (Campbell J. Gibson and Emily Lennon), 1999, "Historical census statistics on the foreign-born population of the United States: 1850–1990" (<http://www.census.gov/population/www/documentation/twps0029/tab22.html>).

US Census Bureau, 2006, "American community survey" (<http://www.census.gov/acs>).

Yokohama, City of, 2006, "Creative City Yokohama" (http://www.city.yokohama.jp/me/keiei/kaikou/souzou/en/index.html)

Index

Accra 144
Akron (Ohio) 83
Amsterdam 18, 19–20, 25
Antoniszoon, Cornelis 25
Asian Tigers 46, 49
Atlanta (Georgia) 83, 91, 93

Baltimore 48, 49
Bangalore 57, 162–5, 166–8
Bank of Amsterdam 19
Beijing 154
Bell, D. 61
Berlin 109, 142
Berry, B.J.L. 66
Birmingham (Alabama) 34
Birmingham (UK) 38
Borchert, J.R. 66
Braudel, F. 14, 27
Braun, Georg 25
Bruges 17–18, 48
Brussels 55
Buenos Aires 123, 142, 154
Buffalo (New York) 83
Burnham, Daniel 35, 36–7
business service 69, 73

capital cities, colonial legacy 142; and
    growth pole theory 145; and national
    economic development 141–4, 146; and
    nationalism 142–4, 146; and pursuit of
    world city status 146, 149; seven types of
    141–2; and world's twenty tallest
    buildings 146, 147
cartography 25–6
Castells, M. 7, 103, 129
Chennai (Madras) 22–3
Chiang Mai (Thailand) 145
Chicago 109
child labor 35

China 13–14, 38, 41, 42, 43, 57, 140–1, 162
Chistaller, W. 14–15
Cincinnati (Ohio) 71–2, 84–6, 103;
    Over-the-Rhine area 3–4
cities, diseconomies of 5; effect of
    technological innovation on 5;
    global–local framework 6;
    growth/decline 2–4, 8; and increasing
    returns concept 2–3; and infrastructural
    dualization 159; literary portrayals 4–5;
    and networked technologies 158; new
    solutions for 87–114; relationship with
    economies 1–2, 8–9; and splintering of
    urbanism 158; and vibrancy of
    downtown business districts 113
City Beautiful Movement 37
Civitates Orbis Terrarum 25
class consciousness 33–4
Cleveland (Ohio) 71–2, 84–6
colonialism 20–1, 23–5, 120, 142
Columbian Exhibition (Chicago, 1893) 36
Columbus (Ohio) 84–7
command and control centers 24–5, 118–19
cosmopolitan city 70–2
Communist Manifesto 32
creative center 108
creative economy 104; and city preferences
    107–8; concept 104–5; and creative
    capital theory 105–9; and creative class
    105–6, 108, 112–31; and creative talent
    104; and developing environment for
    108–9; and economic growth 105–6; and
    economy-culture relationship 113
CreativeTampaBay (Florida) 106
cultural economy 98; and cultural turn 98–9;
    economic dimensions 99–102; industrial
    aspect 99–100; and path dependence
    100–2; and tourism 100; and urban
    redevelopment/gentrification 102–4

De Soto, H. 130
Defoe, D. 20
deindustrialization 40, 98
Denver (Colorado) 88
Detroit 27, 40, 41, 48
developing countries, effect of globalization on 64, *see also* third world cities
Dickens, C. 4
Dutch East India Company 19, 21
Dutch West india Company 21, 24

East India Company *see* Dutch East India Company
Engels, F. 32, 33
ethnic enclaves 165, 168–72

female employment 56
Florence 28
Ford, Henry 37
Fordist city 37–8
Frank, A.G. 6, 122
Friedmann, J. 6, 66, 68, 74, 78, 120, 140

garden cities 35
gated community 134,160–2
Geddes, P. 66
Ghana 144, 146
ghost towns 49
Glasgow 102
global shift, concept 43; and increase in international trade 45–6; and mobility of capital 45; Nike 43–5; and reconstruction/reterritorialization of working class 44–5
global urban hierarchy 74–8
global urban networks 20–1, 23–5, 74–8
global urbanism 54, 94–6
globalization, and consumption patterns 94; contradictory/uneven nature of 157; and convergence 94; definition 156; and dispersal of economic activities 73; effects of 64; and globalization-urbanization nexus 74, 95; and growth of service industries 54–5, 57–60; and homogenization of the world 93–4; impact on third world cities 122–3; and inequalities 156–60; as patchy/incomplete 46; urban adaptation to 60; and urban economic change 7–9; urban impact of 80–1, *see also* world cities
Globalization and World Cities (CaWC) Study Group and Network 55, 75–8

globalizing city 93–6
good governance 87
Gottmann, J. 143
growth pole theory 144, 145

Hadid, Zaha 103
Hall, P. 31, 32, 37, 47, 66, 77, 124, 141
Hanseatic League 16–17, 18
Harlem (New York) 100
Haussman, Baron 35–6
Havana 123
high-tech enclaves 162–5, 166–8
Hogenberg, Frans 25
Hong Kong 49, 54, 55, 146, 152
Howard, Ebenezer 35
Hull House (Chicago) 34–5

Immigrants' Protective League 35
India 66–8, 120, 121, 162–5
industrial cities 3–4; and Britain 29–30; class/race 33–4; Fordist/mass production 37–8; and global shift 43–6; and increase in international trade 45–6; locational theories 31–2; and manufacturing decline 38–40; and mechanization 33; planned 34–7; and political power 33–4; and postfordist city 40–2; and production innovation 28–9; and social conflict 32–4
Industrial Revolution 28–30
informal economy 129–131
information technology (IT) 162–5
International Labor Organization (ILO) 130
International Monetary Fund (IMF) 140, 157
Istanbul 122, 154

Jacobs, J. 3, 9, 105
Johannesburg 123, 148–9
Juvenile Protective Association 35
Juvenile Psychopathic Clinic 35

Kafka, F. 4
Kennedy, P. 19
Keynesian-New Deal 38, 60
knowledge-based economy 55, 104
Kondratieff cycles 29, 37
Kuala Lumpur 146, 154
Kumasi 144

Lagos 132
Lautensack, Hans 25
Le Corbusier 35
Letchworth garden city 35

London 55, 74, 90, 117, 124, 142
Los Angeles 100, 101
Losch, A. 31

Madras (Chennai) 22–3
Magellan, Ferdinand 13
Manchester 30–2, 109
manufacturing 38–40; and flexible
    production 40–1, 45
Marx, Karl (Marxism) 32–3
mega capital cities 143
mega cities, black holes in 134; decoding
    the chaos of 132; and environmental
    concerns 131; and government policy
    130–1, 133; growth/development of
    124–6, 133–4; and informal sector
    129–31; level of immigration in
    132; location of 127; and national/
    global companies in 134; and
    population-economic change 126–34;
    and slums 126, 129; typology of 126,
    128; and urban infrastructure 126
mercantile cities 3; cartographic
    representations 25–6; and colonialism
    20–1, 23–5; as command/control sites for
    capitalism 24–5; global reach 20–1,
    23–6; historical development 14–20;
    main European trading networks 16–20;
    and maritime trade 13–14; and municipal
    government 24; northward shift in power
    19–20; and public culture 24;
    range/threshold of goods/services 14–15;
    transcontinental archipelago of 15–16;
    and world trade/urban development link
    16
meta cities 126
Mexico City 142, 154
middle class 39
Mumford, L. 3

Naisbitt, J. 5
Napoleon III 36
Native Americans 21, 23
Negroponte, N. 5
neo-liberalism 42, 131
new capital project 142–3
new economy 98
New World 19, 20–1, 23–4
New York 49, 55, 74, 90, 100, 101, 117, 142
Nike 43–5, 46, 140

offshore financial center 162
Ohio 71–2, 83, 84–7, 103

Olympic Games 93, 141, 152–3, 154, 172
Osaka 102, 110–11
overseas call center 162

Paris 36, 55, 100, 142
path dependence 100–1
planned cities 34; new models for 356; and
    social activism 34–5; and urban
    interventions 36–7
population distribution policy 143–4
post-colonial urbanism 8
postfordist city 40–2
primate cities 120, 124, 139, 142, 171;
    development of 134–5; economic effects
    of 136–7; environmental concerns 137;
    examples of 135; location of 135–6
producer service 50–6
public sector 56–7

race 34
Radburn garden city (New Jersey) 35
Renaissance 15
retailing 42, 45
Romantic Movement 30

Saminiati merchant trading company 16
San Francisco 92
San Jose (California) 41
satellite-metropole relations model 6, 122–3
Schenectady (New York) 41, 49
Seoul 149–55, 165, 168–72
service industries 42, 73; concept 50; and
    female employment 56; financial 54;
    global shift of 57–60; and globalization
    54–5; growth of 50, 54–6; and polarized
    job market 55–6; and advanced business
    and producer services 50, 54–5, 69, 73;
    and role of public sector 56–7; shift from
    manufacturing to 49, 56, 58
Silicon Valley 41, 166
Singapore 55
splintering urbanism 158–9
South Korea 38, 43–4, 46, 143–4, 149–55
Southey, R. 30
Soweto 148
Spain 19, 142
stick-and-carrot measure 133
"Strategies for Creative Spaces and Cities
    Project" (Toronto, Canada) 109, 112
suburbs 35

technology 5, 7, 158; and high-tech enclaves
    162–5, 166–8

Tehran 123
third world cities 6–7, 115–16, 140, 156;
and capitalist world system 122; Cold
War background 117–18; colonial cities
120; and digital divide 158–9; early
writings on 119–20; and ethnic enclaves
165, 168–72; and gated communities
160, 161–2; graduating from 146,
149–55; and high-tech enclaves 162–5;
impact of globalization on 122–3; and
income/wealth 159–62; and inequality in
globalizing 157–60; in Urban Studies
118–24; and the media 157; mega cities
in developing world 124–7, 129–34;
post-colonial/nationalist aspects
143–4, 146; and primate cities 120, 124,
139; and quality of life 137–9; and
satellite-metropole relations approach
122; and spatial modeling 121–2; as
under-theorized/under-emphasized
118–19; and urban problems 121; in
urban studies 118–24; Western approach
to 120–1
Thompson, E.P. 33
Tokyo 74, 117, 142, 152, 153
Toledo 102
Toronto 109, 112
trade liberalization 45–6
transnational networks 95
Treasure Fleet 13–14
Trinidad 161–2

United Nations Human Settlements
Programme (UN-HABITAT) 8, 124,
126, 133, 137, 160
urban competitiveness 81, 82, 87–9, 158;
and economic development 82, 83–7,
88; and good governance 87; and
policy initiatives 88–9; and
psychological edge 89; and social
development 87; and workforce
development 87
urban economic development 88–9
urban economy, and deindustrialization 7;

and globalization 7–9; and
suburbanization 7
urban growth, and decline 84–5; importance
of size 48–9; and ratchet effect 48–9;
three-stage model 49; twenty largest
cities in US (1850–2000) 51–3
urban policies 81
urban regimes 58–9; adaptation to
globalization 60; central/local 59;
instrumental 59; and neo-liberal agenda
60; organic 59; and regime change
59–60; symbolic 59
urban success story 88

Venice 18–19, 48

Wal-Mart 42, 95
'wannabe' world city 94
Weber, A. 31
Williams, R. 5
working class 44–5
world cities 6–7, 8, 9, 156;
American/cosmopolitan 70–2; as
command/control centers 72–4;
commonalities of 74, 80–1; comparative
studies 94–6; concept 66; criteria 66–7;
criticisms of thesis 69, 72; data on 75–8;
and dynamic urban processes 95; and
economic change 69; emergent 157; and
geographic bias 72; global perspective
65–6, 68–9; globalizing processes 81,
94–6; hierarchy of 67, 72, 74–8; and
media comments 92–3; politics of status
89–93; servicing of 73–4; shared
characteristics 80; and social polarization
157; in urban studies 66–72
world city network 77
world city school 69
world city status 89–93
World Cup Finals 149, 153–4
world urbanism 119

Yokohama 110–1
Yoruba 120